Historic Tales *of* FLATHEAD LAKE

BUTCH LARCOMBE

Published by The History Press
Charleston, SC
www.historypress.com

Front cover, bottom: Seli'š Ksanka Qlispe' Dam, formerly named Kerr Dam, and the Flathead River below the lake. *Courtesy of Dolan Pobran.*

First published 2024

Manufactured in the United States

ISBN 9781467154741

Library of Congress Control Number: 2024931901

CONTENTS

ACKNOWLEDGEMENTS

The roots of this book run deep, in some cases tapping into vague stories shared decades ago by family members. Other elements stem from tidbits passed along more recently. New or old, each of them came to these pages with the help of many people.

Earlier versions of a few of these stories initially appeared in print elsewhere. I owe deep thanks to *Montana Quarterly* magazine and Scott McMillion for taking the publishing bait. Similarly lured in were Myers Reece, Justin Franz, Tristan Scott and Hunter D'Antuono at the *Flathead Beacon* and *Flathead Living*.

There are lots of individuals who contributed time, insight and photographs to this project. Hopefully, this list includes most of them: Denny Kellogg, Ed Gillenwater and Kyle Stetler of Bigfork; Tom Bansak and a number of others at the Flathead Lake Biological Station; Karen Dunwell of Polson; Mark Fritch, the photo archives curator at the Mansfield Library in Missoula; Dave Ruby and others at the Somers Company Town Project; Jim Atkinson and others at the Northwest Montana History Museum in Kalispell; Kate Sheridan and the Flathead Lakers and a former Lakers employee, Hilary Devlin; Leslie Kehoe and Kehoe's Agate Shop near Bigfork; Patrick Jones at Bay Books & Prints in Bigfork; Dennis O. Jones, also of Bigfork; Ruth White in Dayton; Jasmine Morton of Kalispell; Rick Weaver of Kalispell and Eric Hanson of Lakeside; Emily von Jentzen of Kalispell; and Edie Cope in Missoula.

Deep and special thanks to my oldest sister, Pam Larcombe Coffman, who provided invaluable editing on this project. Thanks also to Artie Crisp and the crew at The History Press for believing that this book idea had merit. Finally, thanks to Jane Larcombe, my wife of nearly four decades.

INTRODUCTION

In August 1959, Montana senator Mike Mansfield, in the midst of a legendary congressional career, sent a letter to U.S. president Dwight D. Eisenhower. The subject? Flathead Lake.

Mansfield wrote:

> *Mr. President, in western Montana we have what I consider to be the finest and most beautiful natural freshwater lake on the North American continent, Flathead Lake. This lake is one of the scenic wonders of the West, it abounds with recreation potential. It's a sportsman's paradise. Along its shores you will find homes and summer cottages that are comparable to those anywhere in the Nation. Flathead Lake is in the heart of a fertile valley. This lake plays an extremely important role in the generation of power and the control of flood waters in the Northwest.*

The 1959 letter was an explanation of legislation recently introduced by Mansfield and James A. Murray, Montana's other U.S. senator, seeking to make the maximum and minimum levels that had been in place on the lake for decades a part of federal law. The goal was to put an end to the periodic proposals to transform the lake, formed eons ago by a massive glacier, into an unpredictable reservoir, its water level dramatically raised and lowered on a bureaucratic whim.

The letter was not Mansfield's first on the topic. In his first year in the U.S. House of Representatives in 1943, he wrote to President Franklin

D. Roosevelt concerning a wartime proposal to raise Kerr Dam and use Flathead Lake as a storage reservoir. Not given to overstatement or flowery language, Mansfield told Roosevelt the letter was the "most important letter I have ever written in my life" and that the great swings in the lake level "would make a stinking morass of the most beautiful scenic area in the United States."

Flathead Lake holds centuries of history, much of it shared in an oral tradition dating back centuries. No one truly knows how long Indigenous people have lived near or used the lake. But the stories told with pictographs that stretch from the head to the foot of the lake are evidence of a spiritual presence or the desire for spiritual guidance that could stretch back a thousand years or more.

The density of the occupation sites around Flathead Lake and along the Flathead River, wrote Carling Malouf, a University of Montana anthropologist, "indicates that this was, perhaps, the most important center of ancient life in Montana west of the Continental Divide."

In more recent times, the Salish, Pend d'Oreille and Kootenai have used the lake, at times camping for extended periods on its shores. While members of each tribe moved seasonably and often ventured east of the mountains in search of bison, the area on or near Flathead Lake or the rivers connected to it was a traditional base. Natives hunted deer, elk and bighorn sheep; fished in streams and lakes for bull trout, westslope cutthroat and mountain whitefish, along with other fish species; and gathered camas, bitterroot and many varieties of berries.

The arrival of non-Native explorers in the early 1800s brought significant change to the areas near Flathead Lake. The Treaty of Hellgate in 1855 created the Flathead Indian Reservation, which included the southern half of the lake. The reservation, the treaty said, would be for the "exclusive use and benefit" of the tribes, a promise broken repeatedly in ensuing decades.

The construction of the Northern Pacific Railway across the reservation in the 1880s, the allotment of tribal lands to individuals, the sales of "surplus" allotments, the opening of the reservation to homesteading, the creation of a large irrigation system to promote agriculture and the construction of a large hydroelectric dam on the lower Flathead River (formerly known as the Pend d'Oreille River) in the 1930s all reshaped reservation and Native life.

The surge of White settlers arriving by train at Ravalli and heading north to Polson fueled the steamboat era on Flathead Lake, which lasted about twenty-five years. The boats, carrying people and cargo, spurred the

A Kootenai man fishes from a sturgeon-nosed canoe. R.H. McKay photo. *Archives and Special Collections, University of Montana.*

growth of communities around the lake, including Polson, Bigfork, Dayton and Lakeside. The arrival of the expanding Great Northern Railway in the upper Flathead Valley in the early 1890s, accompanied by its voracious appetite for rail ties, drove the development of Somers as one of the largest lumber operations in the region as well as an early hub where steamboats met steam engines and railcars.

While steamboats eventually gave way to roads and automobiles, Somers, a top-to-bottom company town owned largely by the Great Northern Railway, remained an industrial hub for almost 50 years, cranking out rail ties and paychecks for hundreds of workers. Even though the lumber operation is long gone, the town's centerpiece hilltop mansion remains, its walls the guardians of more than 120 years of stories.

Flathead Lake holds its share of secrets. Alongside the puzzles presented by pictographs, its waters harbor the remains of sunken vessels and, possibly, very large fish or even a monstrous creature. The lake is the place of human tragedy in the form of construction deaths, plane crashes and biological mayhem caused by tiny shrimp.

Homesteaders and others arrive by train at Ravalli on their way to the Flathead. H. Schnitzmeyer photo. *Denny Kellogg collection.*

The lake is also a place of triumph. Despite significant population growth and the environmental consequences that accompany it, the water in the lake remains remarkably clean and clear, a credit to the vigilance of the people who hold it dear. It's a place that invites a swim, whether it's a quick dip in a sheltered bay or a stiff open-water test of human will.

The Salish name for the Upper Pend d'Oreille, translated to English, is "the people living along the shore of the broad water," the broad water being Flathead Lake. As Pend d'Oreille elder Mitch Smallsalmon noted in 1977, the big lake is a place where Native people enjoyed a bountiful life for centuries. Today, like the Pend d'Oreille, Kootenai and those who came before them, all who live near or use the lake have, as Smallsalmon put it, been made "wealthy from the water."

PART I

The Early Years

CHAPTER 1

THE SPLENDID WRITINGS

On a pleasant, still day in mid-August 1904, Morton J. Elrod, a biology professor at the University of Montana, and two other men boarded the *Big Fork*, a small steamer, at Dayton in Flathead Lake's Big Arm and headed east.

In a paper published in 1908, Elrod, a man of broad interest and curiosity, offered this description of the excursion: "On the way out as the boat rounds the point and steams out into the open lake the passage is so close to the cliff on the shore, and the hieroglyphics are plainly seen by the passengers with the naked eye." More specifically, the hieroglyphics are what Elrod later described as a "splendid series of Indian writings." The scientist, who founded the university's biological station in Bigfork in 1899, called the site the Pictured Rocks.

A century later, on similarly pleasant summer afternoons, it's not uncommon to spot a flotilla of boats bobbing in the lake, the passengers peering at the cliff where the splendid writings remain remarkably visible. As Elrod did in 1904, the modern-day visitors take pictures of the site now known as Painted Rocks.

That day in 1904, Elrod and his companions managed to land the *Big Fork* among the rocks below the cliff and went up for a closer inspection. It being late in the day, they didn't stay long. After camping overnight on Wild Horse Island, to the south, they returned the next day; Elrod took detailed notes and made photographs using a bulky camera and glass plates. Another

The cliffs at Painted Rocks on the lake's west shore. H. Schnitzmeyer photo. *Denny Kellogg collection.*

man measured the depth of the water, concluding that the cliff continued well below the water surface and noting that just 50 yards offshore, the lake was 157 feet deep.

The men also broke off several pieces of the shale bearing the writings and took them to Missoula for further study, a practice viewed as unethical and possibly illegal by latter-day archaeologists, anthropologists and others who study Native rock art.

Even with no public access by land, due to the boating audiences, the Painted Rocks remain the most easily visited of the seven rock-art sites along or near Flathead Lake. While they are likely viewed by hundreds of people each year, the Painted Rocks are still a source of deep mystery. What do the pictographs depict? Who painted them? How old are they?

There are no clear answers. Writing in 1950, Carling Malouf and Thain White, the west shore rancher, museum operator and amateur archaeologist, recounted conversations with Baptiste Mathias, a Kootenai leader. Mathias relayed a story told to him by his father, who was born in 1826. Guardian spirits, known as *nupeeka*, were in a canoe on the lake and anticipated the imminent arrival of a great flood. "They knew too that this place was going to be taken over by people later." As the flood took place, the guardian spirits were able to reach a cliff at the edge of the lake and escape the rising water. In a meeting atop the cliff, the spirits made a plan to communicate with those expected to come later. The spirits decided to put their "signatures" on the rock and share how they intended to help those new arrivals.

"The symbols are names and the lines show how long they were there. One hundred years ago, maybe one thousand years ago," Mathias said, according to the researchers' account. "When the white people came there were diseases, and the Indians that got them went to these places to seek nupeeka (or the powers they believed those spirits possessed). They put their names down there and how many days they were there.…It shows that they got their power or medicine there."

While Malouf doubted that the pictographs at the Painted Rocks were made after the arrival of Whites, the idea that the images there and elsewhere in western Montana are the work of those who sought spiritual power, visions or "hunting magic" has won acceptance among anthropologists and others.

In his initial visit in 1904, and on a return trip in 1907, Elrod recounted finding two hundred distinct markings at the Painted Rocks. The pictographs, he wrote, depicted at least fifteen different animals, including five buffalo

A Native man, possibly a Salish leader known as Sam Resurrection, at Painted Rocks. R.H. McKay photo. *Archives and Special Collections, University of Montana.*

with distinctive humps, deer and moose. There were also geometric figures and numerous tally marks, which are common at many western Montana rock art sites.

Like the Painted Rocks pictographs, other examples of rock art around Flathead Lake are found in relatively inaccessible spots, including on cliffs, on lakeshore rocks and on islands. At Painted Rocks, most of the pictographs and markings are found on two large slabs of rock about one hundred yards apart on the large cliff. Like those at most other sites, the markings at Painted Rocks were painted in red. There is evidence that at least one painting that's partially chipped away was replaced by a grouping of dots and lines at a later date.

But determining how long ago any of the pictographs were painted is challenging. Anthropologist James Keyser, who was born in Ronan and earned bachelor's and master's degrees in anthropology at the University of Montana before getting his PhD in that discipline at the University of Oregon, says most of the rock art in western Montana is likely two to three thousand years old. Keyser believes many of the pictographs and the much rarer petroglyphs in the Flathead area are similar to those found to the west in a region known as the Columbia Plateau, an area he has studied extensively over the decades.

Animal figures, tally marks and other drawings at Painted Rocks. Hunter D'Antuono photo. *Flathead Beacon.*

Determining the age of Columbia Plateau rock art "is *very* difficult," he told this author in 2023. Researchers look to the images themselves to find clues that could hint at their age. Does the rock art depict weapons? Works with guns are relatively new, while spears hint at likely prehistoric art created before the use of the bow and arrow.

While there are pictographs in western Montana, including some west of Kalispell and near the south end of Flathead Lake, that include images of horses, those at Painted Rocks do not. Experts believe that horses didn't arrive in Montana until after 1700. "Painted Rocks," Keyser said, "probably dates between two hundred and one thousand years ago, with the majority of the images less than five hundred years old, but that's just a best guess."

There is also uncertainty about who painted on the rocks. While the Kootenai have had a continual presence in the area over the last 150 to 200 years, the Pend d'Oreille also camped on and used the area around the lake, as did other earlier Native inhabitants.

For many modern visitors, the mystery of their age and authorship is part of the allure of the "splendid writings" that are nearly as visible to the naked eye as they were to Elrod in 1904.

CHAPTER 2

THE LAKE'S BIG BOATS

In the early days of January 1910, a steam-powered freighter named the *Big Fork* left Somers on a late afternoon, bound for Polson. It carried a heavy load of cement, five passengers and a crew of four. The freighter and its cargo never made it to Polson.

While details are scarce, news accounts of the *Big Fork*'s fate agree that the steamboat ran aground along the lake's west shore, mostly likely in the dark. The *Missoulian* reported that the boat "came to grief on hidden rocks off Angel Point, with the complete loss of its cargo." A more in-depth account from the *Daily Inter Lake* on January 3, 1910, had the boat finding rocks farther down the lake:

> *The wind and warm wave of the preceding day had entirely cleared the lake of ice and no difficulty was experienced until the vessel had gone about twenty miles down the lake, and it had grown dark. Off Painted Rocks a severe storm was encountered, and in a short time, the vessel drove heavily on the rocks, the cargo shifting under the force of the blow and the high seas, and the next wave swept over the deck, sinking the boat.*

The hull of the boat was about ten to twelve feet underwater, according to the unattributed newspaper account, while its pilothouse remained visible above the lake's surface. While luggage was swept away, the passengers and crew made it safely ashore, although the vague news story offered no details of the escape.

FREIGHT STEAMER BIG FORK SINKS

The steamer Big Fork, a freight vessel plying on Flathead lake, was wrecked in the vicinity of Painted Rocks or Wild Horse island Saturday night and went to the bottom. The passengers and crew reached the shore in safety.

An article from the *Whitefish Pilot* about the sinking of the *Big Fork* steamer on the lake's west shore. *Somers Company Town Project.*

The sinking of the *Big Fork* is one of many colorful events that occurred during the period that big boats plied the waters of Flathead Lake, carrying passengers and all manner of freight, including livestock and grain. The boats were born of necessity, given the lack of roads on either side of the lake in the early days of White settlement. While travelers could travel north by train from Missoula to Ravalli and, later, into the upper Flathead Valley, going around the lake on foot, on horseback or in a wagon was arduous at best.

Noting the upper Flathead was one of the last areas of Montana to be settled by non-Native inhabitants, early steamboats "provided a practical means of transportation to this virtually uninhabited area," wrote Dr. A.B. Braunberger, a Kalispell optometrist who in 1965 completed a research paper on the Flathead Lake steamboat era based on decades of personal research and interviews with many who had direct knowledge of the boats and their operation.

While many of the early commercial boats were powered by steam, the first to carry passengers was likely a sailboat named the *Swan.* Built in Polson, the thirty-foot boat battled unpredictable wind and river currents on its way up the lake to Dooley's Landing, a few miles up the Flathead River above the lake. The trip typically took a week. The *Swan* was later sold and its sails swapped for a boiler. Renamed the *U.S. Grant*, the retrofitted craft, launched in the mid-1880s, is believed to have been the first steamboat to operate on the lake.

A photo by Herman Schnitzmeyer captured the *Klondike*, one of the largest boats to operate on the lake, arriving in Polson. *Denny Kellogg collection.*

In the coming quarter century, dozens more steam-powered boats would traverse the lake, often ferrying freight and passengers between Polson and Somers, with stops along the way. Some of the big flat-bottomed boats would make their way up the Flathead River to Demersville, one of the earliest White settlements in the valley above the lake.

The *Crescent*, a flat-bottomed sternwheeler, made its way up the winding, often treacherous river to Columbia Falls, likely trying to establish a water link between the Northern Pacific Railway in Ravalli and the Great Northern Railway in the upper Flathead Valley.

Braunberger notes that while the arrival of the Great Northern in the early 1890s offered a faster and more convenient way to reach the upper Flathead, the use of steamboats on the lake likely peaked in about 1915, when more than twenty boats carried passengers and freight between various points, typically between March and late December.

One of the best-known boats to operate on the lake was the *Klondike* (also the *Klondyke* for a time), a steam-powered sternwheeler that enjoyed multiple

lives. The first version, constructed by Eugene Hodge in 1909, offered direct service between Polson and Demersville, a trip that typically took about six hours. The second version of the boat, known as the *New Klondike* and launched in 1910, could carry 425 passengers and 110 tons of freight. The workhorse, at 120 feet in length and with a beam of 26 feet, was one of the largest vessels used on the lake. Photos of the boat depict its imposing presence on the lake and how its shallow draft allowed it to land in places, such as Yellow Bay, likely inaccessible to deeper-hulled vessels.

While steamboats, and the smaller, faster gasoline- and diesel-powered boats that came later, provided a key transportation link before the development of passable roads and the proliferation of the automobile, the romance of the era was tempered by the mercurial nature of the big lake. While it was generally favorable to navigation, Braunberger wrote that experienced lake navigators never underestimated "its suppressed yet violent potential, the submerged truculence, that existed within this body of water." Sudden storms, he wrote, "could produce a distasteful combination of wind and water."

The *Big Fork* was hardly the only big boat to find trouble. Less than a year into operation, in September 1887, the steamer *Pocahontas*, shortly after passing through the group of small islands that form the "Narrows" north of Polson, encountered a storm that brought large waves crashing over its deck. The boat's pilot, apparently seeking calmer waters, tried to pass between the lakeshore and Melita Island and struck submerged rocks. The boat sank within minutes. A woman and child were placed in a lifeboat, while the male passengers and crew swam ashore.

In December 1909, the *Queen*, a steam-powered tugboat, was trying to make its way through Polson Bay when the crew discovered a hole in the hull, likely caused by ice. The boat, which carried a number of passengers, turned toward shore in an attempt to reach safety, but after water doused the boat's wood-fired boiler, it sank about forty feet from dry land. All the passengers and crew made it ashore, although some vowed to "not again embark upon the lake until the blue birds nest again in the springtime," according to an account in the *Lake Shore Sentinel*, a newspaper of the era.

There were many more mishaps that resulted in sunken boats, although details of many of the incidents are elusive. A largely hand-drawn map of the lake pinpointing many landing spots that dotted its shores was created in 1969 by Thain White, an amateur historian and proprietor of the Flathead Lake Lookout Museum. The map also noted the approximate locations where a number of the big boats sank over the decades, including several incidents in Polson, at least two in Somers Bay and one near the entrance to Bigfork Bay.

One of the sunken boats mapped by White was the *Kee-O-Mee*, a fifty-four-foot pleasure boat built for John Sherman and Bert Saling, co-owners of the Flathead Motor Sales Co. in Kalispell. Built at a cost of $12,000 in 1928, the boat had four staterooms, a kitchen, a bathtub and room for plenty of guests.

"Our family would have lavish weekend parties on the boat," recalled Dorothy McGlenn, Sherman's youngest daughter, in an interview with the *Flathead Beacon* newspaper in 2016. "My father and Bert would invite all the local bankers, lawyers and of course the Kalispell City Council for a day on the water. Everyone was dressed in their finest white shirts, ties and hats."

The owners installed a new diesel engine in the *Kee-O-Mee* in the spring of 1937. Early in an excursion on Somers Bay, the engine caught fire and the flames spread. Unable to subdue the flames, the boat's owners rowed away in a small boat. According to one account, a tugboat towed the burning boat out into the bay to keep the flames from spreading to other boats and the dock. Within an hour, the boat sank.

Several versions of the *Klondike* hauled people and freight on Flathead Lake. *Somers Company Town Project.*

Nearly eight decades later, using a remotely operated vehicle fitted with underwater imaging equipment, a Kalispell man, Kyren Zimmerman, was able to locate the remains of the *Kee-O-Mee.* A team, including Zimmerman, dove to the boat, which sits on the lake bottom in about seventy feet of water. Visible were the boat's anchor and propellor and other items, including a teapot and plates. "What's left of the ship is very well preserved," Zimmerman told a *Beacon* reporter at the time. "It's really like stepping back in time down there."

Some of the items from the *Kee-O-Mee* found a new home in the Northwest Montana History Museum in Kalispell. About six years after finding the sunken boat in Somers Bay, Zimmerman teamed up with the Bigfork Art and Cultural Center and a small group of history researchers on a project to locate other sunken boats and sites of cultural merit in the lake. The ongoing work is intended to catalog and preserve the sites, which could include locations important to Native people.

Speaking to a group in Bigfork in 2023, Zimmerman and others talked of plans to use technology to develop underwater maps in key areas, including

The *Kee-O-Mee* was used by its car-dealer owners to entertain customers and others before it sank in Somers Bay. *Northwest Montana History Museum.*

the lake's more heavily used bays, to possibly find sunken boats, remnants of railroad activity and, likely, interesting objects currently unknown. "We've got a significant number of boats that are still submerged in the lake that we are excited to go survey," Zimmerman told the group. "What we are not aware of is all the mysteries that are below in Flathead Lake."

While the golden era of big boats on the lake had largely ended by 1930, other large vessels continued to travel the lake. The SS *Hodge*, a one-hundred-foot, sixty-five-ton sternwheeler fitted with pieces of the *Klondike*, worked for decades on the lake building docks and other structures. There have been several large tour boats, including the *Retta Mary*, a sixty-four-foot, double-decked craft, later rechristened the *Far West.* Originally based in Polson and later in Lakeside, the big boat has operated on the lake each summer, with a few exceptions, since 1971.

Another boat, the MS *Flathead*, was based in Polson and hauled thousands of passengers on lake cruises. The sixty-five-foot boat originally operated in Glacier National Park as the *St. Mary* before being trucked to Flathead, where it was renamed and refitted. With its white hull and red trim, *Flathead Courier* editor Paul Fugleberg wrote, "it was the unquestioned queen of the boating set" for more than half a decade.

But the boat, owned and piloted by Stanley Koss, met a colorful end just prior to the summer tour season in 1961. The boat was in the Narrows at the

Scenic Cruiser Burns on Lake

The *Flathead*, a popular tour boat, was severely damaged by fire in the Narrows and towed to Polson, where it sat along the lake for several years. *Daily Inter Lake.*

head of Polson Bay when an engine fuel line apparently broke, spewing diesel onto a hot exhaust pipe. The fuel ignited and flames spread rapidly, prompting Koss to beach the burning boat as he and a teenage passenger, Steve Reum, escaped unscathed. But the MS *Flathead* burned to the waterline, and its charred remains were later towed to Polson to a spot near the current Kwataqnuk resort, where it sat rotting for several years.

Fugleberg noted that the MS *Flathead* was more than just a tour boat to many. "She was to Flathead Lake residents today what the *Klondike* and other famous lake ships were to people of this area back around the turn of the century," Fugleberg wrote. "She was the biggest, best known and most popular of the ships on the lake."

Chapter 3
THE DURABLE *HELENA*

While the names and stories of many of the big steamboats that operated on Flathead Lake have slipped into the thick fog of memory, reminders of the *Helena* can be found almost a century since its last lake run.

The hull of the *Helena* sits in the Flathead River near what was the community of Holt, visible at times when the river is low. The venerable Kehoe's Agate Shop, the only commercial structure still in Holt, was built during the Great Depression with beams, planks and even nails salvaged from the steamboat. Between the shop and river sit the *Helena*'s pilothouse, rusting metal propellers and an anchor, along with a worn wooden rudder.

While many other steamboats burned or sank, some more than once, the *Helena*, built in Bigfork in 1915–16 after many of its predecessors had disappeared, survived at least in piecemeal fashion due to its durable construction and the passion of the Kehoe family.

James Kehoe, a marine engineer with extensive experience with steamboats on the Great Lakes, came west driven by a plan. "He wanted to have a steamboat of his own, and he knew he couldn't do it in the Great Lakes," said his granddaughter Leslie Kehoe in a 2023 interview. He had a big boat in the style of the big Midwest cargo steamers in mind. He considered other lakes in the Northwest, but after visiting Flathead, he looked no further. "The Flathead was so beautiful, and he just loved it."

With his father-in-law, Leonard Robinson, an experienced tugboat captain, and other relatives and helpers, Kehoe started work on the boat in 1915. Assembled near the mouth of the Swan River, the craft that would later be

Pieces of the *Helena* rest near Kehoe's Agate Shop at Holt, once the site of a Flathead River ferry crossing. *Author's collection.*

named for Montana's capital city was constructed with planking purchased from the Somers Lumber Company, while its hull and keel were formed from larch, chosen for its tendency to harden when wet. Key components included a boiler and a salvaged marine engine shipped from Chicago.

The *Helena* was constructed with a belt of sheet iron around the waterline that helped protect the hull from ice and logs. The boat, 110 feet long with a 25-foot beam, got a trial run in March 1916 when it left Bigfork and broke a channel through ice up to three feet thick to the mouth of the Flathead River. The roughly three-mile trip took five days.

The elder Kehoe's son, James J. "Jack" Kehoe, recounted the memorable maiden voyage to an interviewer decades later: "We would back off and take a run at the ice and make a length or two each time, but it was slow work. We burned wood in the boiler and it took a lot of it bucking ice."

The *Helena*, with a higher bow than earlier Flathead Lake steamers, had four holds that were loaded with a derrick and boom. While a couple of other boats that operated in the early 1920s, including the *Klondike*, carried large loads of passengers and freight, the *Helena* was primarily a cargo workhorse that hauled bulk wheat, apples, lumber, potatoes and baled hay between various points on the lake. Jack Kehoe, who began piloting the *Helena* at age fourteen, also recalled one particular load: 110 cords of wood ferried from

The steamboat *Helena*, one of the last to operate on the lake, featured a bow design used on the Great Lakes. *Northwest Montana History Museum.*

Bigfork to a site on Hellroaring Creek near Polson, where the wood was used to fire a steam plant that generated electricity.

Along with ice, the wind on the lake could be a challenge, particularly on the north end of the lake, where a storm could bring waves eight to ten feet high. The sturdy *Helena*, Jack Kehoe wrote, "was the only boat ever built on the lake that could go out in any weather or bad storms, and get through without laying over or much loss of time."

While waves or ice couldn't thwart the *Helena*, the completion of roads around the lake and the proliferation of trucks eventually did. The boat was dismantled in 1931–32, its hull left to rest parallel to the east bank of the Flathead River.

Jack Kehoe opened the agate shop on Holt Drive in 1932, after working with a Flathead Valley jeweler. He learned to harvest and cut rock and later built a saw that used blades with diamond tips. The industrious Kehoe, who as a youth rented a fleet of rowboats to visitors and anglers for one dollar per day and, later, operated a tugboat used in the construction of Fort Peck Dam in northeast Montana during the 1930s, worked in the rock shop for decades. "He cut Montana agate for more than fifty years," said his daughter.

Jack Kehoe, who helped operate the *Helena* (*background*) with his father, at the oars of one of his rental rowboats. *Leslie Kehoe collection.*

Pieces of the wooden hull are still sometimes visible in the winter months, although some were washed down the river and into the lake during the massive flood of 1964. Others have been displaced by the waves and erosion caused by increased recreational boat traffic on the river above the lake.

Generations of the Kehoe family have lived near the banks of the river since 1922. "When the bow of the boat was still there, we used to go play on it and dive into the river from it in the summertime," Leslie Kehoe recalled. "There is still a little bit left, but not a lot. That's too bad."

Along with boxes of handwritten remembrances and photographs, she also has a logbook from the *Helena* that documents the trips and loads, its pages yellowed by the decades. The handwritten entries offer a glimpse into life on the boat and a colorful time on Flathead Lake.

"Steam boating was some of the hardest kind of work and fighting storms, and currents in high water, running long hours, and taking chances around reefs and bars in the dark were part of a day's work, but I wouldn't have missed the experience for anything," Jack Kehoe wrote in 1956.

CHAPTER 4
FROM PATHS TO PAVEMENT

While travelers from afar could reach the upper Flathead Valley via a Great Northern Railway passenger train starting in the early 1890s and were able to venture north from Missoula to Ravalli on the Northern Pacific even earlier, the stretch between the rail points often involved horses, wagons, a boat or a very long walk.

"Flathead Lake itself served as both an obstacle and as the means of traveling for traveling north and south," area historian Henry Elwood noted in his history *Kalispell, Montana and the Upper Flathead Valley*. "Early settlers in the region generally found travel extremely difficult."

On the west shore, the first route likely followed what is known to some as the Old Indian Trail, a footpath that stretched at least from the Elmo-Dayton area north and up a long ridge that passed west of Rollins. The crest of the trail was known as Medicine Rock Pass and traversed what some described as the Big Lodge Divide. From there, the trail ventured closer to the lakeshore for a distance as it headed toward the area that later became Lakeside and to points farther north.

Segments of the trail, while overgrown and obscured in many spots by logging and road-building activity, have been located by history buffs over the years. In a 2011 note to an acquaintance, Doug Johns, whose family were among the first White settlers in the area, described the trail as a resembling a trough, ten to twelve inches deep in places. "Once you were in it, it was very easy to follow," Johns wrote. "Another defining characteristic is that it doesn't wander but is exceptionally straight."

Traveling along the lake's west shore could be perilous. Photographer and date unknown. *Denny Kellogg collection.*

Area historian Thain White wrote that the trail up and across the Big Lodge Divide was used by some of the first wagons and White settlers as they made their way to the valley north of the lake. By all accounts, it was a tough trip.

Sam Johns, an early homesteader and relative of Doug's, made this observation about the climb from the south to Medicine Rock Pass: "The drop off south was for some distance very steep, and you wonder how they ever got a wagon up to the top."

Frank Tetrault, in recounting his family's passage into the area in 1882 when he was ten or eleven years of age, confirmed the difficulty of the wagon travel. "We had a rather hard time getting to the summit—we had to unload our stuff from the wagons and carry it to the top."

While historians believe the Old Indian Trail eventually became part of what was known as the Wagon Road to Demersville, the early riverside settlement above the lake, the steep pass was hardly the only obstacle. Ahead, south of Lakeside, sits Angel Hill, where early wagon travelers recalled not only the climb up but also the perilous descent. In many instances, ropes were attached to the rear of wagons to allow people and even horses to slow the wagons on the narrow downhill passage.

Described as a "road," the route was unusable in the winter and barely passable the remainder of the year. The Wagon Road to Demersville was

A stagecoach on the road to Somers. J.S. Thompson photo, date unknown. *Denny Kellogg collection.*

abandoned in the early 1900s. Flathead County built a road to the east of the wagon route and made a series of improvements and route changes until US Highway 93 was finished in the area in 1929 or 1930.

On the other side of Flathead Lake, a road of any sort along the east shore didn't come until more than thirty years after a passable route was created on the west shore. The steep slopes of the Mission Mountains came down to the edge of the lake in many spots, leaving no natural path for a road. Travel along the east shore for many years involved a horse or a person's two feet. Some early inhabitants recall that a trip from the Bigfork area to Yellow Bay could take up to two days due to the steep, rocky terrain.

It took an enterprising prison warden and the strong backs of inmates to make a road a reality. A statewide movement to improve roads and accommodate increasing automobile travel took root in the first decade of the twentieth century. Nationwide, the use of inmates on construction projects was increasingly common, and in Montana, the idea of using convict labor gained steam. "The chief aid of the state to the betterment of our highways can come only by working the convicts on the roads," Montana governor Edwin Norris proclaimed in 1910.

The so-called good roads movement found support among other state officials and garnered the enthusiastic backing of Frank Conley, who oversaw the operation of the state-owned prison in Deer Lodge, part of which had been built with convict labor.

In March 1912, a crew of forty to fifty prisoners began work on what became known as the East Shore Road. The project was launched as part of an agreement between state and prison officials and Flathead County. The goal was to construct a road with a gravel surface, twenty to twenty-four feet wide, with drainage ditches on both sides. Flathead County (Lake County was not formed until 1923) agreed to provide equipment, including shovels, picks, horses, wagons and scrapers. Area businesses pitched in money for horse feed and blasting powder.

The prisoners were paid fifty cents a day for grueling manual labor. While explosives and pneumatic drills were used to break up the plentiful rock, removing the rock and creating the roadbed involved horses, strong backs and hand tools. "It was hard work on those convict gangs, but there were perks that came with it," noted Jon Axline, a historian for the Montana Department of Transportation, in a 2022 interview.

The perks on the east shore project included good meals, lodging in tents near Flathead Lake and, often, evening musical entertainment. But the biggest benefit might have come in the form of being free from the poor

Montana State Prison convicts working with rudimentary tools helped construct much of the first road along the lake's steep east shore. Date unknown. *Montana Historical Society, 947-981.*

conditions at the prison in Deer Lodge. The facility was overcrowded and lacked indoor plumbing for many years.

"I think it was a privilege to get out of there," Axline said. Most of those who worked on the projects were doing time for theft, burglary or cattle rustling rather than more violent crimes. Surprisingly, the guards who oversaw the convict laborers and camps were unarmed. On the East Shore project, the convict crew numbered as many as 111 prisoners, who toiled over three construction seasons. There were just thirteen escape attempts, the small number possibly a reflection of the punishment—prompt shipment back to the Deer Lodge prison.

Conley, the prison official, spoke about the use of convict labor in glowing terms: "There in the freedom of the mountains, the petty criminal develops brain and brawn....From the brow of the burglar and bank robber drops the sweat of honest toil."

Conley may have had more than prisoner well-being and rehabilitation in mind with his strong support of convict labor. While the prison was owned by the state, Conley and a partner operated it under a contract for a number of years. When deals to use convict labor were reached, at least a portion of the money landed in the prison operators' pockets.

Later, Conley became a state highway commissioner in addition to his prison role. When picking convict labor projects, "he decided who got to use them and who didn't," Axline said. "There was a definite conflict of interest in that Conley was probably making quite a bit of money off convict labor."

After the end of World War I, former military construction equipment was made available to states. Conley, as chairman of the state highway commission, had control of equipment and labor on road projects.

"He had his own little empire for sure, not only in Deer Lodge but in Helena," Axline said, noting there are mixed views of Conley's historical legacy. "He was a complicated guy, and either you loved him or hated him."

While early progress on the East Shore Road was steady, a disagreement between Flathead County and state officials about who should pay the guards stalled the project in 1913. The dispute ended with the state paying the guards.

The use of convict labor found controversy in Montana from its earliest days. Organized labor, a powerful force in parts of the state for many decades,

The road along the lake's east shore was not fully paved until the 1940s. *Montana Historical Society, 946-828.*

long opposed the practice. As the demand for better roads grew, private highway contractors seeking work from the state and local governments were also critical. The prison labor program ended in 1925, partially due to the powerful opposition.

More than 230 miles of state roads were built during the roughly twelve years the program was in place, including pieces of the state's main highways. But the 27 miles of the East Shore Road completed in 1914 at a cost to Flathead County of just under $32,000 represent the largest project completed with convict labor in Montana.

The road retained its gravel surface until 1938, when the stretch from Bigfork to Yellow Bay was paved. The remaining link to Polson was not completely paved until 1949; the current Montana Highway 35, with its deep dips, twisting turns, road-hugging trees and flirtations with the lakeshore, largely follows the tortuous route tamed by convicts more than a century ago.

CHAPTER 5

A COMPANY TOWN

Located at the northern tip of Flathead Lake, with a landmark water tower, a main street lined with century-old buildings, a colorful bar, a fine little museum and a big yellow mansion on a hill, Somers is clearly a town with a story.

Longtime residents can pinpoint the former main street locations of a doctor's office, a bakery and the building that has housed a bank, a grocery store and now a café. Just steps away from the café sits Sliters Lumber and Building Supply, the venerable hardware and lumber emporium. The store, at its current site since 1933, bears the name of a pioneering family that has cut a wide swath through the history of the Flathead Valley. The first store at that spot was the "company store," started by John O'Brien, who built the lumber mill that was the central force in early-day Somers.

For at least fifty years, Somers was a top-to-bottom company town, created to feed the Great Northern Railway's voracious appetite for wooden rail ties. Typically, each mile of new line needed three thousand wooden ties, and the miles of need grew quickly as the Great Northern dug and pounded its way to the Pacific Northwest.

In the late 1890s, the Great Northern's founder, James J. Hill, dispatched O'Brien, a successful lumberman, to Montana's timber-rich Flathead Valley to find a location for a lumber mill to produce the ties. For $10,000, O'Brien bought 350 acres at the head of Flathead Lake from the McGovern family. The railroad built a spur line south from Kalispell, forging a link to the main Great Northern east–west line. At the same time, the lakeside

An early-day view of the main street of Somers and the company store. *Somers Company Town Project.*

lumber operation took shape, fed by towboats that hauled large rafts of logs harvested in the region's thick forests across Flathead Lake to the new mill.

The first ties were cut in 1901, and the town, named for George O. Somers, a Great Northern manager, boomed. Within two years, the mill was in full operation, at first cranking out triangular rail ties, the shape intended to be cheaper to produce and simpler to place than rectangular ties. (Within a few years, the weaknesses of triangular ties became evident, and the mill began producing rectangular ties.) Needing employees, the mill offered would-be workers in Minnesota the chance at a future in remote, wild Montana for a $12.50 one-way rail fare.

The workers came to the remote valley, not only from Minnesota but also from afar: Italy, Germany and the Scandinavian countries. They settled in small houses—the company built 122 homes on the fledgling townsite—fed by company-produced water and electricity. Rent for the homes was one dollar per year, water was a buck a month and electricity was billed at the rate of fifty cents per outlet. A hotel, churches, saloons, a school, a pool hall and a barbershop popped up. Workers and their families settled along narrow streets that would acquire names like Swede Hill, Happy Hollow, Pavilion Hill, Slab Alley and Battle Hollow.

Men on a large log boom. Morton J. Elrod photo. *Archives and Special Collections, University of Montana.*

Small steamboats, including (*left to right*) the *A. Guthrie*, the *Kootenai* and the *Willis*, moved logs in Somers Bay. *Somers Company Town Project.*

An overview of Somers in 1906, including the fence that some believe was intended to keep labor agitators away. *Somers Company Town Project.*

An area known as Pickleville, populated largely by Italian families, was named for its proximity to the "pickling" plant, where ties were treated with zinc chloride and, later, a creosote-oil mixture. In 1910, Somers had 650 residents. In the coming decades, it would grow to more than 1,100.

In its early years of operation, the big mill in the little town, where workers put in ten-hour shifts for pay ranging from twenty to sixty cents per hour, saw a number of instances of labor unrest. Three strikes in first decade of the twentieth century produced vitriol and shutdowns of plant operations for varying periods. At the core of the unrest were the "Wobblies," members of the Industrial Workers of the World (IWW), the union that had established a firm foothold in the lumber mills, logging camps and mines of the Pacific Northwest and elsewhere.

During a strike in the summer of 1909, the union painted a grim picture of the little town and its owners, the Great Northern Railway and James J. Hill. "The newest approach to hell on Earth is Somers, Mt," wrote F.W. Heslewood in the *Industrial Worker*, a newspaper published by the IWW. "A company town, company store, company doctor, company wood, company water, company light, company house, company bank, company roads, company post office, a lot of company suckers called scabs and a company high board fence around hell. God hates this place so much that he blew the roof off the Roman Catholic church." (The roof was damaged by lightning.)

Life in Somers eventually stabilized as the mill and tie plant cranked out hundreds of thousands of railroad ties and other wood products. At its peak in 1937, the Somers Lumber Co. employed 375 workers and produced sixty million board feet of wood products. "This town, in its heyday, was the industrial center of the Flathead Valley," noted Dave Ruby, president of the Somers Company Town Project, in a 2018 interview. "It was the largest sawmill in the state at one point."

A retired railroad engineer, Ruby was born and raised in Whitefish and later moved to Bigfork. But both his parents were born in Somers. One grandfather drove one of the squat S-2 locomotives that hauled loads of ties around the mill yard and treatment plant. Another grandfather worked as a brakeman and later as a "tiebucker," loading heavy ties, by hand, on and off small railcars as they moved around the plant. Ruby, pointing to the small locomotive outside the museum, noted, "We had countless hours as kids riding around on these things." Later, in his own rail career, he made runs from Whitefish to Somers at the helm of a locomotive that hauled ties to the railroad's main line.

The Great Northern closed the mill operation in 1948, citing a dwindling supply of timber. A salvage company bought pieces of the mill, the homes,

Dave Ruby's grandfather drove an S-2 locomotive around the lumber mill yard in Somers. *Author's collection.*

the water system and the power plant. The homes were sold individually, in some cases to the families that had rented them.

Another mill, the DeVoe Lumber Company, operated on a smaller scale in Somers for about ten years after the closure of the Somers Lumber Company. An early-morning fire in June 1957 severely damaged the DeVoe operation and a large amount of lumber. DeVoe never resumed operation, marking the end of lumber milling in Somers.

The tie plant treated ties hauled to Somers by rail until 1986. The plant was tabbed an EPA Superfund site in 1984. Almost four decades later, monitoring wells watch for signs of creosote and other industrial chemicals used to preserve wooden ties that could migrate toward Flathead Lake.

The Somers Company Town Project was formed in the early 2000s to capture and preserve the town's history. The group raised more than $140,000 and rounded up many hours of volunteer labor to build a small museum in the town's center, starting with little more than a small batch of old photos. Local folks—many of whom worked in the lumber mill or tie plant or were related to someone who did—donated many items linked to the town's early industrial operation and social life. The donations were fueled by a desire to preserve the unique history of the colorful little town at the head of Flathead Lake.

CHAPTER 6

THE MANSION ON THE HILL

While history is plentiful in Somers, the onetime company town at the head of Flathead Lake, one of its most colorful chapters swirls around the nearly eight-thousand-square-foot yellow mansion that sits on six acres atop a hill in the center of town. Within its walls reside stories of opulent living, reclusive residents and financial intrigue.

Possibly designed by Billings architect Joseph Gibson, the mansion—with twenty rooms, including a ballroom and many ornate flourishes—was built for John O'Brien, who founded the lumber operation in Somers with the encouragement and financial backing of Great Northern Railway baron James J. Hill. Railroad ties, a principal product of the mill, were a key component in Hill's empire building.

Along with fourteen bedrooms and only three bathrooms, the mansion also included a multiple fireplaces, grand windows and, nearby, a carriage house. Originally named Alta Vista, it was completed in 1903.

Flathead Valley author and researcher Jaix Chaix, writing in the *Flathead Beacon* in 2014, portrayed the mansion as a monument to O'Brien's ambition and desire for status, not only in the fledging community but also in the burgeoning Northwest lumber industry. He wrote: "Situated atop the hill, O'Brien could preside over the company and town in every direction....He could watch over the doings at the dock, along the rails, and around the shops, all the same. Hence, the mansion—much like the man himself—was intended to be omnipresent, ever keeping a watchful eye above the company town."

The mansion built by John O'Brien was completed in 1903 and has led a colorful life. *Somers Company Town Project.*

Accounts from early-day residents say O'Brien; his wife, Anna; and other family members lived mostly on the mansion's second floor. The first floor was used for social gatherings and company business, while the third floor often housed guests and visiting company officials.

Despite its grandness and prominent location, the mansion was home to O'Brien for just a few years. O'Brien sold his interest in the lumber mill to Hill and the Great Northern in 1906 and moved to Vancouver, British Columbia, where he continued in the lumber business, likely until his death in 1914.

The Great Northern bought the mansion and used it for company purposes for several years. Between 1911 and 1915, it was operated as a hotel named the Mountain Inn. After the hotel shuttered, the structure was used for company offices before eventually becoming a residence for several mill managers, including Edward McDevitt, who oversaw what had become the Somers Lumber Company from 1931 to 1948.

McDevitt eventually purchased the home, and several generations of his family lived there for roughly fifty years. In the latter years of the family's ownership, the home and its occupants largely avoided the public eye. The last McDevitt occupant was Edward's son, E.N. "Ned" McDevitt Jr., who worked for the Somers Lumber Company, including two summers on the *Paul Bunyan* towboat that hauled large rafts of logs from around the lake to Somers.

The driveway to the mansion adjoined the white two-story home that belonged to Ruth and Jim Hellen for decades. The Hellens helped take care of the mansion property for a time. Back when their family operated the grocery store in Somers, the Hellen children would deliver one quart of milk each day—no more, no less—to one occupant of the sprawling home. Others recall getting a reluctant reception on Halloween forays to the home. Like the lumber mill and tie plant, the rumor mill about the mansion and its occupants steadily hummed.

Jasmine Morton, who grew up near Somers, recalled her fascination with the home as a child. As an elementary schooler, after spotting a light on in the home, she took a plate of chocolate chip cookies to the door, hoping to get a peek inside. After she knocked on the door, the light was hastily switched off and the door went unanswered. Morton left the cookies on the porch, bewildered but still enchanted by the mysterious mansion.

In 2005, by then more than a century old, surrounded by overgrown shrubbery and guarded by "No trespassing" signs, the home landed on the real estate market. Christin Didier, a Lewistown, Montana native and the winner of the Miss Montana USA pageant in 1997, reportedly paid $1.1 million for the property and took up occupancy.

The mansion on a hill in the center of Somers served as a hotel, the Mountain Inn, for a time. Hess photo. *Somers Company Town Project.*

Just a couple years later, a storm severely damaged the mansion's roof, and not long thereafter, a fire further damaged the structure. Didier vacated the mansion, filed an insurance claim for repairs and sought reimbursement for the cost of replacement housing while repairs were made. Subsequent investigation revealed that Didier had greatly overstated the cost of her rental home, which led to a fraud conviction, eviction, a bankruptcy filing and the mansion, now in foreclosure, sporting another "For sale" sign. The asking price for the mansion and acreage overlooking Somers, the north end of Flathead Lake and a wide swath of the Flathead Valley: $399,900.

A California flight attendant, Christine Manson, apparently entranced by its price and potential, bought the property in 2013. Manson spoke with news reporters and others of her renovation plans. But little work took place, and the mansion, with its badly damaged roof and broken windows, sat largely untouched and continued to deteriorate. Some residents feared it would be demolished and the property sold to a developer. A social media group, Save the Somers Mansion, digitally conveyed those concerns.

In an interview with the *Flathead Beacon* in the months after she bought the hilltop property, Manson admitted to being stunned by the intense interest in the big home. "I never imagined this huge spotlight on me when I bought it. It's overwhelming," she said. "I'm going to do whatever I can

The mansion offers occupants a grand view of Somers and Flathead Lake. Hunter D'Antuono photo. *Flathead Beacon.*

to preserve what I can, but it seems like the whole town thinks it has a say in what happens."

In 2020, roughly seven years after her purchase, Manson put the mansion on the market, asking $890,000. Among the first to get a showing was Jasmine Morton, the girl with the cookies, now a Kalispell attorney, who ended up buying the sprawling home with the large hole in the roof, crumbling plaster and broken windows.

The mansion got a new roof and new windows, a coat of paint, foundation repairs and electrical and plumbing upgrades. Morton's plan included reconfiguring rooms to make suites with adjoining bathrooms and refurbishing public spaces, making the mansion and carriage house inviting for weddings, family gatherings, meetings and possible use as a bed-and-breakfast inn.

Just a few weeks after closing on the purchase, Morton held an open house at the mansion. More than four hundred people showed up. As renovation work progressed, Morton issued several more invitations that drew many who, like the cookie-bearing girl years before, were looking for a peek inside.

In an interview with the author in 2023, Morton offered a simple explanation for the new chapter in the colorful story of the mansion: "It has always been part of the community, and people have been curious for so long. I want to keep it as part of the community."

CHAPTER 7

THE JOURNEY OF THE *PAUL BUNYAN*

It's hard to believe the *Paul Bunyan*, long tucked under a roof outside a funky museum in Polson, was once the most powerful boat to operate on Flathead Lake.

Constructed in Somers in 1926, the tugboat, designed by Seattle naval architect L.E. "Ted" Geary, was outfitted with a 180-horsepower diesel engine and was the flagship of the fleet of workboats used to haul logs to the Somers Lumber Company mill at the head of the lake. The muscular *Paul Bunyan* could tow twice as much as older steam-powered craft such as the *A. Guthrie*, *Howard James* and *Kootenai.* The sixty-five-foot, fifty-ton boat, constructed largely of coastal fir at a cost of $126,000, could tow as much as a million board feet of logs, although its loads were typically not that large.

Al Fischer, who worked for the Somers Lumber Company for fifty-five years, starting at age sixteen, was part of the crew on the *Paul Bunyan*'s first log-towing trip in 1927. He also was aboard for the big boat's last trip in 1947. Sharing memories with local historian and museum curator Dennis O. Jones, Fischer recalled that a trip from Somers to Dayton, roughly two-thirds of the way down the lake, would typically take about three hours aboard the big tug. The return trip, with a large raft of logs trailing behind the boat, would often take thirty-two hours, "if it was good goin'," according to Fisher. The return trip, at average speed of one and a half miles per hour, often included an overnight stop at Table Bay, on the lake's west side.

Once at Somers, the logs that were added to the inventory floated in the bay, awaiting milling. Fischer recalled that one of his duties was to measure

The diesel-powered *Paul Bunyan* towed large log booms from around the lake to Somers for several decades. *Somers Company Town Project.*

the size of the mass of logs at Somers. At one point, he told Jones, the floating logs covered ninety-nine acres of the lake's surface.

There were times when not even the stout tug could navigate in some of the storms that occasionally pound the lake. One year, in the fall, the towboat sought refuge in a bay near Wild Horse Island, Fischer told Jones years later. "We were stranded for 10 days due to large waves."

The big boat couldn't ride out a post–World War II economic slowdown and changes in the Northwest Montana timber industry, including better roads and the proliferation of logging trucks. While the tie-treatment plant at Somers continued to operate until the mid-1980s, the Great Northern Railway determined it was cheaper to buy ties milled elsewhere than operate the Somers plant. The shutdown left workers and the *Paul Bunyan* high and dry. The boat, idled, remained in dry dock at Somers for a number of years.

In 1957, the *Paul Bunyan* found a lofty home and became a visible landmark. The proprietors of the Flathead Lake Lookout Museum, Ernest White and his son, Thain, bought the boat and moved it by truck to the museum perched on a cliff above the lake's Deep Bay. Along US Highway 93 about four miles south of Lakeside, the boat and museum offered a commanding view of the lake and drew about thirty thousand visitors a year, according to news accounts. Thain White added a living room to the boat's deck, made room for a kitchen by cutting away a portion of its hull and lived in the boat with his family for a number of years.

The younger White sold the museum in the early 1970s to the O'Neil Jones family of Bigfork, and Dennis O. Jones operated the museum for a number of years. In 1978, the museum and much of its contents were sold at auction. But the *Paul Bunyan* remained, with time and weather exacting their toll. The then-owner of the property, with plans to develop the thirty-acre site, had no interest in the large, dilapidated boat anchored just off the highway.

Enter Gil and Joanne Mangels, who had founded the Miracle of America Museum in Polson in 1981. The couple bought the boat in 1985, for the reported price of $5,000, and announced plans to move it to Polson. "If we hadn't bought it, they would have burned it," Gil Mangels told *Missoulian* reporter Vince Devlin in 2010. "But they wouldn't give it to us. We had to buy it."

The sixty-five-foot *Paul Bunyan* has made its two most recent voyages over land. *Ruth White collection.*

The fifty-ton towboat *Paul Bunyan* has been anchored at the Miracle of America Museum in Polson since the mid-1980s. *Author photo.*

The *Paul Bunyan*, again towed by a big truck, navigated the curves and hills of Highway 93 for two days before reaching port at the museum on Polson's outskirts in May 1987. The couple patched the hole in the hull, removed the living room and took other steps to restore the *Paul Bunyan* to its original form. For a time, visitors to the museum were allowed to wander the boat's deck.

But the initial plan to restore the *Paul Bunyan* has never materialized. While the museum's operators were able to raise money to construct a cover to better shield it from the elements, further restoration has been financially stymied. After more three decades at the museum, it is plagued by dry rot in its hull and deck.

Gil Mangels has the blueprints for the big boat, obtained by Jones from a family member of a man who worked on the *Paul Bunyan*, along with old photos and memorabilia from its original owner, the Somers Lumber Company. While the *Paul Bunyan* has undisputed historical value, Mangels harbors regret about buying it. "We just felt it was an important part of Flathead Lake history," he told interviewer Devlin. "To be honest, it's been a big money drain for us, but it elicits such joy and interest from so many people."

CHAPTER 8

BUILDING THE DAM

There was likely dancing in the streets of Polson on a Monday night in May 1930. The cause of the frivolity in the little town at the south end of Flathead Lake? A long-awaited, highly contentious decision from a federal agency to grant a fifty-year lease to a subsidiary of the Montana Power Company that would allow the construction of a big dam on the Flathead River about six miles southwest of town.

"Polson had its biggest night in history tonight," began an account in the *Missoulian*, where the lease decision was front-page, banner-headline news. "Every man, woman and child in town joined in the merry making. Bonfires glowed into the night and every kind of noise-making instrument that was available was brought into use."

The dam project, discussed and debated for more than a decade, was also big news across western Montana and especially in Butte, the home of Montana Power and its largest customer, the Anaconda Copper Mining Company. The first power line from the dam would run south to Anaconda and Butte to fuel the copper company's mining and smelting operations. The lease announcement came about six months after the October 1929 stock market crash. The clouds of a looming depression hung over much of the nation, including Polson and the surrounding Flathead Indian Reservation.

The "Polson dam," as it was called in the construction days, would bring jobs to many, including tribal members, and—after initial opposition from the power company—annual rental payments to the tribes for use of the site. The initial rent agreement provided $2.84 million over a twenty-year period.

The 1930 groundbreaking for the dam in the canyon of the Flathead River southwest of Polson. H. Schnitzmeyer, photographer. *Denny Kellogg collection.*

But to some Native residents, the changes to traditional reservation life and the prospect of a big concrete dam in a narrow canyon where water tumbled fiercely were a heavy cost. "For them it was a sacred place," wrote Thompson Smith, a tribal historian, "a place to be respected, a place where human being should be humbled rather than heedlessly exerting the ability to transform the natural world."

The dam project was the latest development in a period of intense change on the reservation. It started with the allotment of parcels of tribal land to individual members; later, nonmembers were allowed to own and homestead on reservation land, and a large irrigation project intended to promote farming was developed. The overall goal, at least in the eyes of some, was to push Native residents into the White world, like it or not. The dam project also brought industrialism and wage labor to people who for generations had hunted, fished and gathered berries, camas and bitterroot for food.

The big dam project "was part of that assimilation era," said Brian Lipscomb, the CEO of Energy Keepers Inc., the tribal corporation that now operates the dam, named Seli'š Ksanka Qlispe' (SKQ), for the Confederated Salish and Kootenai Tribes (CSKT). "The way of life for tribal people was severely impacted, negatively." While there were deep concerns about the project, at least among traditional tribal members, the Native inhabitants at the time lacked the political and economic clout "to effectively say no," Lipscomb said in an interview with this author in 2021. The CSKT did not have a tribal government or self-governing authority until years after work on the dam started.

In the months before the granting of the license to build the dam, there were multiple visits to the reservation by officials of the power company, including president Frank Kerr. Speaking to interviewers decades later, some Native leaders alleged the company paid tribal members to sign a petition supporting their bid to get the federal license for the dam. Others said they saw Kerr hand cash to tribal chiefs to help win their support.

At the dam's groundbreaking ceremony in 1930, Kerr was made an honorary member of the Kootenai tribe. Chief Koostahtah and other tribal leaders in ceremonial dress presented him with a buckskin commemorating the adoption. Kerr's Kootenai name was A-Kalt-Muc-Quait, which translates to "light."

In an interview years after the dam was completed, Lasso Stasso, a Kootenai man born in 1871, said there were plenty of misunderstandings among Native residents about the lease deal with the power company. "This undoubtedly is due to the inability of the chiefs to fully comprehend the

Work on what became Kerr Dam started in 1930 but was halted for a number of years during the Great Depression. Photographer unknown. *NorthWestern Energy collection.*

nature of their agreement," he said. "When our chiefs leased the ground for the dam, they told me I was to get $25 the first year, then $50 the second year and then $75 the third year and so on for 10 years. But it stopped at $50. What happened to the rest? When the dam was built our people were to get rich from it. But we didn't."

While the CSKT won the right to buy the dam—and a $9 million annual rental fee to be paid by Montana Power—in 1985 during a federal relicensing process, ownership didn't come until 2015. The bitterness over the dam's development lingered for more than eight decades, the hard feelings fortified by a series of tragic events.

After the lease was signed in 1930, work on the dam began quickly. But in the spring of 1931, as the Great Depression deepened, Montana Power pulled the plug on the Polson project. The Anaconda copper operations were suspended due to falling demand, and without its biggest customer, Montana Power had little incentive to boost its electricity production. But five years later, the company restarted the dam project, bolstered by $5

The big dam project in a steep canyon provided needed jobs to Native and White workers. Date and photographer unknown. *NorthWestern Energy collection.*

million from the New Deal's Public Works Administration. Construction projects—bridges, roads, public buildings and hydroelectric dams—provided jobs and hope.

The jobs at the Polson dam project paid between forty cents and two dollars per hour. There were plenty of takers. At its peak, the dam project on the Flathead River employed about 1,200 workers, including several hundred tribal members, many of whom welcomed the paycheck. Mary Smallsalmon, a Pend d'Oreille woman, told interviewers in 1988 that her brother helped build the dam. "He made good money at that time," she said. "He bought a wagon with rubber tires, (a) team of horses, harness—good stuff."

As was the case with other large Depression-era dam projects, the jobs came with a price. At the Fort Peck Dam in northeast Montana, fifty-nine workers died in construction accidents, while many others suffered and eventually died from dust-related disease resulting from the construction of tunnels through the big earth-fill dam. To the west in Washington state, work accidents at the Grand Coulee project on the Columbia River claimed seventy-eight lives. The death toll at the much smaller Polson dam project reached fifteen. The majority of those killed were tribal members.

The first death on the project came on February 2, 1931, just a few months before work was stopped. Albert Gerry was working on a tunnel that would divert the river around the dam site when witnesses said he lost his balance and fell into the river. Despite the use of dynamite to try to dislodge rocks that might have trapped Gerry, his body was never found.

Two days into February 1936, a train hauling coal and a work crew to a construction camp near the dam site derailed as it descended a hill. More than twenty workers on the train were able to jump into snowbanks along the tracks and escape serious injury. Conrad Davis, forty-seven, and Jessie Rojas, thirty-nine, apparently unable to jump, were killed, crushed beneath tons of coal that spilled from the train. A locomotive engineer told investigators he tested the train's brakes at the top of the hill, but as it descended, ice on the

Frank Kerr was named an honorary member of the Kootenai tribe by Chief Koostahtah (*to his right*) and other tribal leaders. C. Owen Smithers photo. *NorthWestern Energy collection.*

rails left the brakes unable to slow the train, causing it to leave the tracks. A coroner's inquest concluded that the accident and deaths were unavoidable.

By early 1937, the pressure to complete the dam was intense and work was taking place in three shifts, twenty-four hours per day. A deadly incident on March 3, 1937, made news across the state. "Seven Bodies Under Big Earth Slide in Canyon," read the *Missoulian*'s front-page headline. To the east in Great Falls, home to four dams and the center of Montana Power's electric generation operations, the event described as a "rock avalanche" was of high interest.

With water diverted through a large tunnel, work was underway in the riverbed, where concrete would form the base of the dam. There were about sixty men in the area on the graveyard shift when "without warning and with startling suddenness," according to Polson's *Flathead Courier*, a portion of a tall cliff on the river's north bank broke loose, "crushing to death instantly seven men who were at work in that particular place." Three other men who were working nearby rushed to the scene and were injured by subsequent rockfalls from the cliff.

Investigators estimated that the slide buried the men under at least one thousand tons of rock. Rescue and recovery work was delayed until officials were convinced the rockfalls had ended. Removing the bodies took several days, with some of the boulders so large that they couldn't be moved with a bulldozer. Small charges of blasting powder were used to break up the rock and allow it to be moved by a steam shovel.

Killed in the initial rockfall were Dave Sanchez, Clifford Gendron and Jack Anderson of St. Ignatius, Tony Adams of Evaro, Henry Couture of Arlee, Allen Ross of Polson and Joe St. Germaine of Arlee. One of the injured rescuers, Harold McNeeley, died a few days later. Most of these men were married and had children. All but one of the dead were tribal members, the *Missoulian* reported.

Investigators noted that "scaling work" intended to remove possible hazards on the cliff, which was about 170 feet high, had been completed before the accident. C.H. Tornquist, a superintendent for the construction company, concluded the slide was likely caused by rocks loosened when the winter's frost went out.

Slightly more than a week later, another worker, James "Blackie" Emerson, thirty-four, was killed and two of his co-workers injured when a seven-ton steel beam collapsed in a tunnel leading to the dam's powerhouse. About six months after that deadly March, on September 3, 1937, three more men were killed while digging a trench. An embankment above the trench collapsed and trapped the workers under dirt and rock. A fourth man was partially buried but rescued. The lost lives of Paul Innes, thirty-five, of Ronan; Baptiste Pierre, forty-one, of Arlee; and Joe Mathias, twenty-eight, of Elmo pushed the dam's construction death toll to fifteen. Mathias was the son of Kootenai chief Baptiste Mathias.

Despite death and injury, work on the dam never abated. Less than a year after the last deaths, the dam was complete, and the Polson area again celebrated what was viewed as one of western Montana's largest industrial developments. The dam was "a boon to both the red man and white man of Montana," according to one news account.

Thousands of residents were joined by Montana governor Roy Ayers, U.S. senator Burton K. Wheeler, Salish chief Martin Charlo, Kootenai chief Koostahtah Big Knife and hundreds of tribal members at the dam site on August 6, 1938, to dedicate the dam and formally name it after Kerr, the president of Montana Power. There was a band concert, Native singing and dancing, tours, a barbecue of five young bison, photographs and speeches.

Cornelius "Con" Kelley, the president of Anaconda Copper, hailed the hard work and vision of Kerr and said the tall concrete dam allowed the use of energy "wasted" in the cascades and falls of the river to be used for "industrial purpose and social service." Chief Koostahtah, the last of the hereditary Kootenai chiefs, speaking through an interpreter, noted that the day "marks the time when great motors start to make the juice to make the white people happy and bring the money from the white people to our people."

More than three-quarters of a century after the dedication, the juice and money began to flow in a different direction. In 2015, the CSKT, which had begun saving money three decades earlier to buy the dam, paid NorthWestern Energy $18.3 million and took full ownership. Within a few months, the tribes won federal permission to change the dam's name from Kerr to Seli'š Ksanka Qlispe', reflecting the Native names for the three tribes that now owned it. Lipscomb, the top dam manager for the tribes, described the name change as "an honor to our people, who sacrificed so dearly in both the construction and acquisition of this facility."

CHAPTER 9
THAIN WHITE AND THE INFORMANTS

When Thain White died at age eighty-five in 1999, a newspaper obituary offered a brief summary of the Dayton man's life. By vocation, White was a rancher and businessman. By avocation, he was a naturalist and student of local and regional history. As a younger man, he worked on the construction of the Going-to-the-Sun Road in Glacier National Park. A more detailed account of White's life might also include these descriptors: author, researcher, photographer, map maker, archaeologist, anthropologist, historian, museum operator and tourism promoter.

White is probably most widely remembered as the longtime operator of Flathead Lake Lookout Museum, a fixture along US Highway 93 about four miles south of Lakeside. While the centerpiece of the museum was the *Paul Bunyan*, the sixty-five-foot towboat once operated by the Somers Lumber Company, the remainder of the museum was filled with items that reflected the wide interests of White, who joined with his father, Ernest White, to open the museum in 1949. Also, there were items showcasing the younger White's interest in early Montana history: maps, Native artifacts, tools and many photographs that reflected his wide-ranging historical gathering.

"It was full of stuff," Ruth White, Thain's youngest daughter, recalled in an interview with this author in 2023. "The main room was mostly early Montana artifacts. The other part was mostly Native artifacts. It smelled really good in there because of all the buckskin."

The museum included items from the family's sheep ranch, which operated along the west side of the lake in the Dayton and Rollins area,

Thain White (*center*); his dog, Boob; and his friends Korrine Wiggin (*bow*) and Kenny Wiggin (*stern*) haul sheep from Papoose Island, now known as Cromwell Island. Ernest White photo. *Northwest Montana History Museum.*

with some grazing land leased from the tribal government. The Whites also grazed sheep and, later, cattle on Cromwell Island, known initially as Papoose Island, of which they owned a portion. Livestock was moved to and from the island by boat. The White family sold its last portion of the island in the late 1980s.

While the museum was best known as the home of the *Paul Bunyan*, a promotional brochure claimed that along with "an unexcelled view of Flathead Lake," the roadside attraction contained "a thousand interesting items for young and old."

The Lookout museum was also the base of operations for White's publishing endeavors, in which he shared his research on many topics. One publication in the museum's portfolio titled *Vessels and History of Flathead Lake* contains a roster of the steamboats that operated on the lake and includes a list of licensed captains.

The publication also includes a detailed map of Flathead Lake that locates not only islands and other geographic features but also steamboat landings and pinpoints where some of the boats sank. The map was hand-drawn in 1969 by White, who proclaimed himself to be "the authority" on Flathead Lake steamboats.

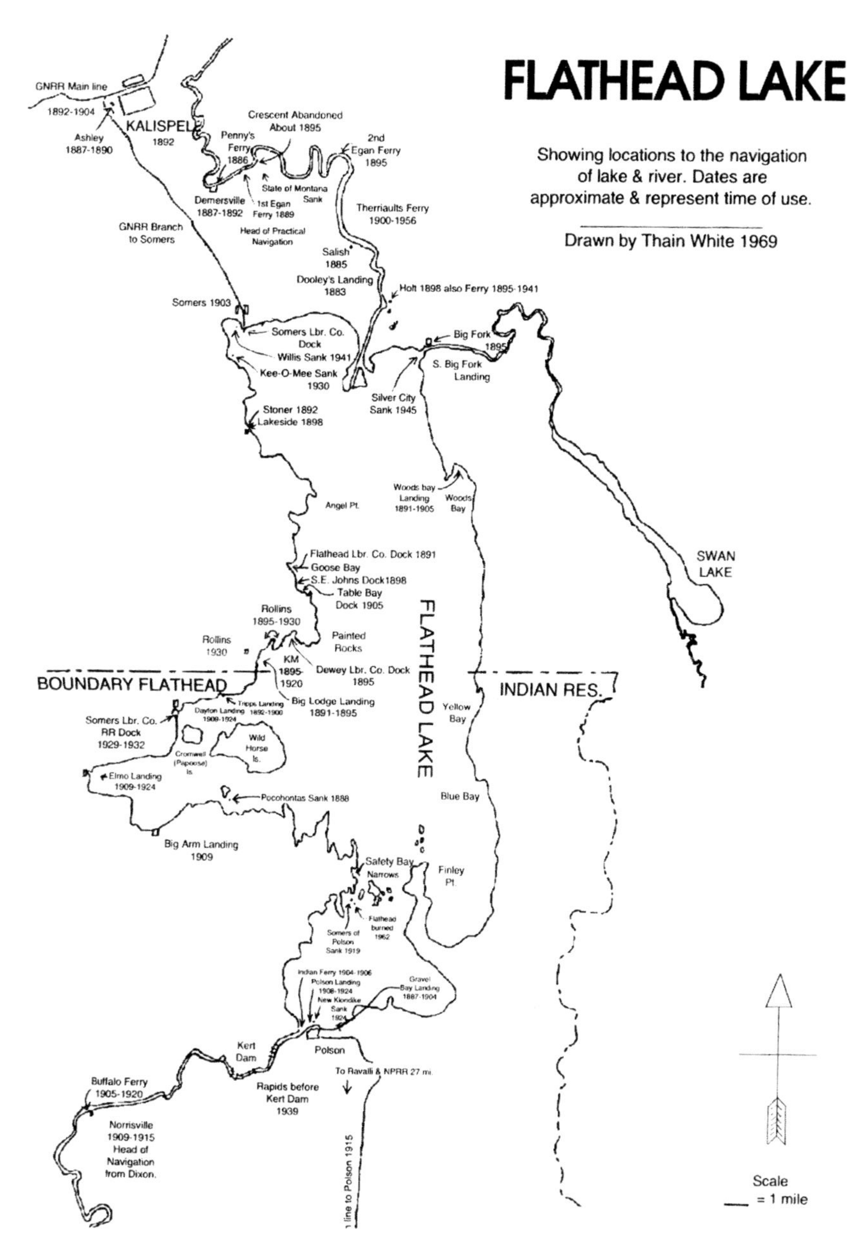

Thain White created this map of Flathead Lake steamboat landings and sinkings in the 1960s. *Author's collection.*

A number of other publications share stories and legends passed along over the years to White by Kootenai inhabitants of the area. Some of the more detailed publications were produced in partnership with Carling Malouf, a longtime professor of anthropology at the University of Montana whose academic career focused on research into a number of Native tribes in the western United States.

In a series of interviews shortly before his death, Malouf shared how and he White often relied on Native people they called "informants" to learn the stories of the people who were the first to live near or use the lake. The "informants" were often folks with firsthand knowledge of events or traditions or who may have learned the stories from close family members.

"He introduced me to some of these older men that really had a knowledge of the older days," Malouf told interviewers in 2004. Many of the sources knew White and his family from the ranching operation. "They were his neighbors, and some of the old-timers became interested in him and he learned from them."

Ruth White said her father had a particularly friendly relationship with Baptiste Mathias, a Kootenai chief, whom he had known since childhood. The time with Mathias and members of his family, she said, fueled her father's deep interest in Native culture. While Malouf had a doctoral degree from Columbia University and was highly regarded during in his forty-year professorial career at the University of Montana, he gave considerable credit to White for his interest in documenting and sharing Native history. "He wasn't tremendously educated," he said of White. "As a matter of fact, he never even graduated from high school, but he had a good-thinking mind. He just amazed me at times what a thinker he was."

Along with his research in the Flathead, White ventured elsewhere in Montana and undertook excavations at both the Big Hole and Bear's Paw battlefields, at Lolo Pass and along the route of the Mullan Road. After White's death, many items related to his research work landed in the collections of several colleges and universities. At the same time, some professional anthropologists and archaeologists have come to question the methods and intent of White's gathering of artifacts and historical items.

The O'Neil Jones family from Bigfork bought the Lookout from White in 1972, and Dennis O. Jones ran the museum for a time. "Thain was just a really interesting guy, to say the least," Jones, himself a Flathead Lake historian, told this author in 2023. "He was very eccentric. He pretended to be ornery, but he wasn't. He had a very interesting sense of humor."

Baptiste Mathias, a Kootenai elder, served as an "informant," sharing stories of early tribal life with Thain White and others. Date and photographer unknown. *Northwest Montana History Museum.*

Jones characterized White's research on steamboats and other topics as a significant contribution to the history of the lake and broader Flathead Valley. "If he hadn't done what he did, it would be gone," Jones said. "History disappears quite rapidly. Thain was here at a good time to capture it. He saw it come; he saw it go."

PART II

The War Years

Chapter 1

THE FLATHEAD LAKE FIGHT OF 1943

Montana contributed mightily to the effort to win World War II, with the deepest contributions coming in the flesh and blood of those who joined up to fight. But a proposal floated in the spring of 1943 brought the war effort to the doorstep of many in the Flathead Valley and across Montana.

The proposal: raise the level of Flathead Lake, at first by seventeen feet and possibly later by an additional twenty feet, by adding to the height of Kerr Dam near Polson. In doing so, the lake would be able to store more water, which could be strategically released to generate more hydroelectricity downstream in the Columbia River in Washington and Oregon.

The proposal from the Bonneville Power Administration (BPA) and the U.S. Army Corps of Engineers and driven by the federal War Department, was said to be critical to the U.S. effort to defeat the Germans and the Japanese. The electricity produced from the dams would bolster the production of aluminum, a key material for making aircraft and other weapons. While the agencies and engineers claimed they had considered other possibilities in the Pacific Northwest, the Flathead Lake proposal was the least expensive, offered the quickest turnaround and, in their estimation, was the "least evil."

At a hearing in Helena on June 1, 1943, an Army Corps official assured those gathered that the Flathead Lake project "was the only solution to the problem of supplying the power needed by the end of 1944 for war production in the northwest."

Flathead Lake, a geological remnant of Glacial Lake Missoula, is the nation's largest natural freshwater lake west of the Mississippi River. The construction of Kerr Dam during the Great Depression by the Montana Power Company affected ten feet of the natural lake's level, allowing it to be drawn down in the fall and winter and filled again in the spring, a regimen that allowed for the generation of electricity and helped maintain a reliable lake level in the warm months, a boon to recreation.

The idea of raising the dam and the lake level brought a tsunami of opposition, not only from the Flathead Valley but also across Montana and beyond. While federal proponents of the proposal saw great potential in adding three million acre-feet of water that could be stored and released as needed, opponents decried the resulting inundation of thousands of acres of fertile farmland in the lower Flathead Valley, plus the flooding of several communities, including portions of Polson, Bigfork, Lakeside and Somers, the latter home to a lumber mill that produced railroad ties for the Great Northern Railway.

Flathead Plan is Called 'Least Evil'

Helena, Mont., June 2—(AP)—Power development studies in the northwest show that Flathead lake in Montana could be raised 17 feet with the least damaging effect on farm lands, says Brig. Gen. Warren T. Hannum of the corps of army engineers.

The Kerr dam-Flathead lake plan was recommended because it would not cause the flooding of as much land as the alternate projects," he told Montana and Idaho state officials and delegations of ranchers, farmers and townspeople.

A proposal to build up Kerr dam near Polson, one of several projects studied by the Bonneville Dam Administration's advisory board, has been protested by ranchers and farmers in the Northwestern Montana area that would be flooded.

General Hannum asserted that raising the lake level 17 feet would boost the storage water by 3,000,000 acre feet. He described the Flathead lake proposal as "the only solution to the problem of supplying the

(Continued on page 8)

The proposal to raise Flathead Lake sparked concern across Montana and the Northwest in the early summer of 1943. *Daily Inter Lake.*

Forrest Rockwood, a Kalispell attorney, joined by T.B. Moore, a local physician, sounded the battle cry in a letter published in the Kalispell *Daily Inter Lake* in late May 1943. They wrote: "We are fighting this war on foreign shores to prevent foreign enemies from destroying our country and way of life, but we believe most of us would prefer being bombed out by the enemy than have our valley destroyed by Washington, Oregon and the Federal bureaus." They added: "If the people don't jump on this with both feet right now then in a year or two there will be a lake between Somers and Whitefish during the spring and a stinking mudflat the rest of the year."

Local residents answered the call, joined by many others from across the state. On June 3, the day of a critical hearing organized by federal officials in Kalispell, many businesses closed, and local residents flocked to Flathead High School hours before the afternoon hearing began. The *Daily Inter Lake* estimated the crowd in the auditorium, in nearby classrooms and gathered

MAKE EVERY PAY DAY BOND DAY

THE DAILY INTER LAKE

LEASED WIRE SE[illegible]E OF THE ASSOCIATED PRESS — EVERY EVENING EXCEPT SUNDAY

VOLUME XXXVI — THE DAILY INTER LAKE, KALISPELL, MONTANA, THURSDAY, JUNE 3, 1943 — NUMBER [illegible]

Huge Crowd Attends Hearing

Striking Coal Miners May Be Ordered to Go Back to Work Monday by President

Reds Shoot Down 123 Out Of 500 Nazi Planes At Kursk

Moscow, June 3—(AP)—A force of about 500 German planes attacked the Russian base at Kursk yesterday afternoon and at least 123 were shot down, the Russian announced [illegible] more, after some planes got through to the city "and haphazardly dropped bombs which inflicted material damage and casualties."

The Russians said they lost 30 [illegible]

Governor Ford Opens The Meeting And Introduces Brigadier General Hannum

A public hearing on the lake plan in June 1943 was described as the largest in Flathead Valley history. *Daily Inter Lake.*

outside, where a radio station set up speakers, at more than three thousand, "undoubtedly the largest crowd ever assembled for a public meeting of any kind" in the Flathead Valley.

Along with statements from many locals, Montana governor Sam Ford shared an extensive list of objections to the proposed lake plan, which he said would inundate fifty-five thousand acres of rich agricultural land at a time when production of crops and cattle was also critical to the war effort. The plan, Ford told federal officials, would destroy towns, homes, businesses and lumber mills and wreak havoc "on the economic structure of the Flathead Valley." The governor also raised a states' rights flag, questioning whether the federal agencies had the authority to implement the plan.

Others were similarly vehement. E.G. Leipheimer, editor of the *Montana Standard* in Butte, said the destruction of the land above the lake "would constitute the most serious economic blow this state and its people have ever suffered." He also noted the importance of respecting human values, "which we regard as superior to such things as horsepower and kilowatts."

The president of the Montana Chamber of Commerce, A.T. Peterson of Billings, accused federal officials of staging a hearing that was little more "than an empty formality," predicting the plan to raise the lake would be put into effect "under the guise of the war emergency."

Some local leaders took a different approach by pitching the idea of building other dams upstream from Flathead Lake, possibly in Bad Rock Canyon near Columbia Falls, in the Glacier View area on the North Fork of the Flathead River and also on the river's South Fork, above the little community of Hungry Horse.

After a second hearing day, federal officials left the Flathead without offering any public comment on the fate of the raise-the-lake proposal. A

Right: A portrait of Mike Mansfield in his first term in the U.S. House of Representatives. Middleton photo, 1942. *Archives and Special Collections, University of Montana.*

Opposite: Public outcry and Mansfield's appeal to U.S. president Franklin. D. Roosevelt helped kill the raise-the-lake proposal. Montana Standard, *July 1943.*

Missoulian story noted the conclusion of the hearing, describing the event as "one of the most memorable gatherings in the history of Montana."

While he and other members of Montana's congressional delegation didn't attend the Kalispell hearing, U.S. representative Mike Mansfield outlined his strong opposition to the Flathead plan in a telegram to the *Daily Inter Lake* just ahead of the event. "I am opposed to destroying a fertile valley and valuable section of our state for the purpose of building up potential resources for Washington and Oregon. I want to see Montana's resources developed for Montana," he wrote, vowing to fight the lake expansion "to the last ounce of my ability and energy."

In early July, Mansfield, just a year into his first term in the House, outlined the controversy in a letter to President Franklin D. Roosevelt, telling him that "this is the most important letter I have ever written in my life." Along with outlining the impacts of the proposed Flathead Lake action, including the loss of farmland and the inundation of parts of several towns, he told FDR that "it would make a stinking morass of the most beautiful scenic area in the United States."

In the wake of the letter, FDR asked Secretary of Interior Harold Ickes to investigate the lake proposal. Just a couple of weeks later, at another

F.D.R. Interested in Flathead Case

Calls for Report From Ickes

At the instance of Congressman Mike Mansfield of Montana's First Congressional district, President Roosevelt has taken a personal interest in the Flathead Lake controversy which raged in Montana during the spring and early summer.

Mr. Mansfield's activities resulted in the compiling of an official record to the effect that no action will be taken by the federal government or any of its agencies to disturb the present level of Flathead Lake.

meeting in Kalispell, top officials from the BPA and the Corps of Engineers met with an audience that included Mansfield, U.S. senator Burton K. Wheeler and Ford, the Montana governor. The news was welcome: the agencies had decided that objections raised in response to the proposed project had a sound basis in fact and "that all plans for altering the level of the waters of Flathead Lake had been definitely and finally abandoned."

Instead, federal officials said, they would recommend the immediate development of water storage in a deep canyon on the South Fork of the Flathead River, well above Flathead Lake. That proposal, also billed as a war emergency, found strong support from local, state and federal officials, including Mansfield.

Despite the alleged emergency, nearly a year passed before Congress approved funding for what became the Hungry Horse Dam project, which would not be finished until 1952. Not long after the war ended, it became apparent that the urgency behind the Flathead Lake proposal was driven by more than a need for wartime aluminum.

The added electric generation from raising the lake "was almost certainly headed for the giant plutonium works at Hanford, Washington, which produced the fissionable material for the atomic bomb," wrote Mansfield biographer Don Oberdorfer.

While Mansfield told Oberdorfer he was unaware in 1943 of the likely use of the electricity that could be produced, his support for the big lake and understanding of its role in western Montana was unwavering. "Even if he had known this, he told me, it would have made no difference in his attitude," the author said in 2003.

Oberdorfer's book *Senator Mansfield: The Extraordinary Life of a Great American Statesman and Diplomat* was based on extensive interviews with the former Butte miner who worked as a University of Montana history professor before entering politics. When asked about the greatest accomplishment of his thirty-four-year congressional career, which spanned not only World War II but also the Korean War, Vietnam and Watergate, Mansfield offered a simple reply: "Saving Flathead Lake."

CHAPTER 2

WAR COMES TO THE LAKE

One cold day in December 1944, Oscar and Owen Hill, father and son, were cutting wood in the Truman Creek area west of Lakeside when they discovered what first appeared to be a cream-colored parachute. The object, later determined to be a balloon, featured Japanese writing and a rising sun symbol. The Hills notified Flathead County sheriff Duncan McCarthy, who gathered the balloon the next day and stored it in a garage. The sheriff also summoned federal authorities, who came to investigate.

News of the balloon discovery understandably sparked curiosity and fueled the local rumor mill. Some accounts say that more than five hundred area residents were able to view the balloon stored in the garage. The first local news report of the find came on the front page of the *Western News*, the weekly newspaper in Libby, on December 14, 1944. The source of the news was a local mail carrier, who had heard it from a Kalispell-based mail carrier. According to this hearsay account, the large balloon didn't have a basket "but in its place was a bomb which had failed to explode when the balloon came to earth." The story delivered by the mail carriers also noted "that secret service men were working on the case." The Libby newspaper reported, "It has been impossible to obtain further confirmation of the story."

The federal authorities, who were actually with the FBI, were tight-lipped. And shortly after the discovery of the apparent balloon bomb west of Lakeside, the federal Office of War Censorship, established after the

Jap Balloon Found In Timber

The first report of a Flathead balloon bomb came in the *Western News*, a weekly newspaper in Libby, in December 1944. *Western News.*

1941 attack on Pearl Harbor, asked the news media to not publish accounts of the balloon bombs, claiming sharing such news would help the Japanese.

In the Truman Creek area, authorities recovered the balloon's envelope, pieces of its rigging and some apparatus. Officials estimated that the balloon had been manufactured in late October 1944 and launched from Japan between November 11 and 25, 1944. That Lakeside-area balloon was just the second found in the continental United States. The first came when a balloon bomb exploded about a week earlier northwest of Thermopolis, Wyoming.

A few months after the discovery near Lakeside on the lake's west shore, in February 1945, what was believed to be a balloon bomb was spotted drifting east of Bigfork, and paper fragments from another suspected bomb were found floating in Flathead Lake that same month. A month later, paper fragments from a balloon were discovered near Coram, east of Columbia Falls.

An Associated Press account of the balloon discovery appeared in several newspapers across the United States, but for months, even as remnants of additional balloon bombs were found not only in Northwest Montana but also across the state and nation, there was very little public mention of the incidents.

Even in the Flathead Valley, where hundreds had seen the balloon bomb found in the Truman Creek area, talk was sparse. L.D. Spafford, the publisher of the *Daily Inter Lake*, offered this explanation to an interviewer: "Everybody was mighty interested, but when the Federal Bureau of Investigation warned not to discuss it, the whole town clammed up. There are too many people with sons and husbands in the service to take a chance on perhaps giving valuable information to the enemy."

This voluntary wartime censorship extended beyond the balloon bombs. Newspapers and radio stations cautiously shared news of the war, much of which was supplied by the federal government itself. It was well after the end of the war in 1945 that federal and military officials declassified documents that revealed the extent of the Japanese use of balloon bombs. It is estimated that Japan launched about 9,300 balloons, all about thirty-three feet wide, each carrying about one hundred pounds of bombs, incendiary

The balloon discovery west of Flathead Lake brought FBI agents to investigate. *Montana Historical Society, Pac 93-01.*

devices and ballast. Of that number, about three hundred were sighted or found in North America. In Montana, thirty-two balloon bombs have been discovered. In some cases, the balloons were largely intact, while in others, only small pieces were found.

Historians believe that Japanese schoolchildren helped manufacture many of the balloon envelopes, which were made of layers of lightweight paper sealed together with waxlike substances. The unmanned balloons, filled with hydrogen, had no propulsion system, instead relying on the narrow band of westerly winds commonly known as the jet stream to carry them across the Pacific at altitudes of twenty thousand to forty thousand feet and at speeds of more than two hundred miles per hour. Aviation expert and former Smithsonian official Robert Mikesh, who did extensive research into the development and use of the balloon bombs, described them as "the first successful intercontinental weapon." Some believe the incendiary devices on the balloons were intended to start forest fires in the Pacific Northwest, which could prompt the U.S. military to shift manpower to fight the fires and possibly impede the development of weapons. Another likely motive behind the balloons? The desire to spark fear.

Of the balloons launched in a roughly six-month period in late 1944 and early 1945, only about three hundred are confirmed to have reached North America, with findings of balloons or related fragments stretching across the United States from Alaska to Michigan and Canada from British Columbia to Manitoba. Montana trails only Oregon in the number of known balloon bomb incidents in the United States.

Launched in the cold, wet months of November and December, the airborne weapons riding the shape-shifting jet stream are believed to have started just two small brush fires and might have caused a temporary power outage at a nuclear plant near Hanford, Washington, where work on components for the nuclear bomb was taking place.

But on May 5, 1945, an incident in Southwest Oregon altered the historical course of the balloon bomb saga. A group of two adults and five children on a Sunday school outing came across what was believed to be a balloon bomb. Moments later, the device exploded, killing the five children and one of the adults. The six deaths are believed to be the only World War II–related civilian deaths on the U.S. mainland.

It was later learned that the Japanese had stopped launching their balloon weapons about a month before the Oregon deaths. Almost two weeks after the deadly explosion, the U.S. government issued a vague statement about the event and warned people of "the danger of tampering with strange objects found in the woods."

While Mikesh and other historians later noted that the voluntary censorship of news about the balloon weapons made it difficult to warn the public about the explosive objects, government officials defended the practice, saying the lack of news coverage about the impact of the balloon bombs "had baffled the Japanese, annoyed and humiliated them, and has been an important contribution to security."

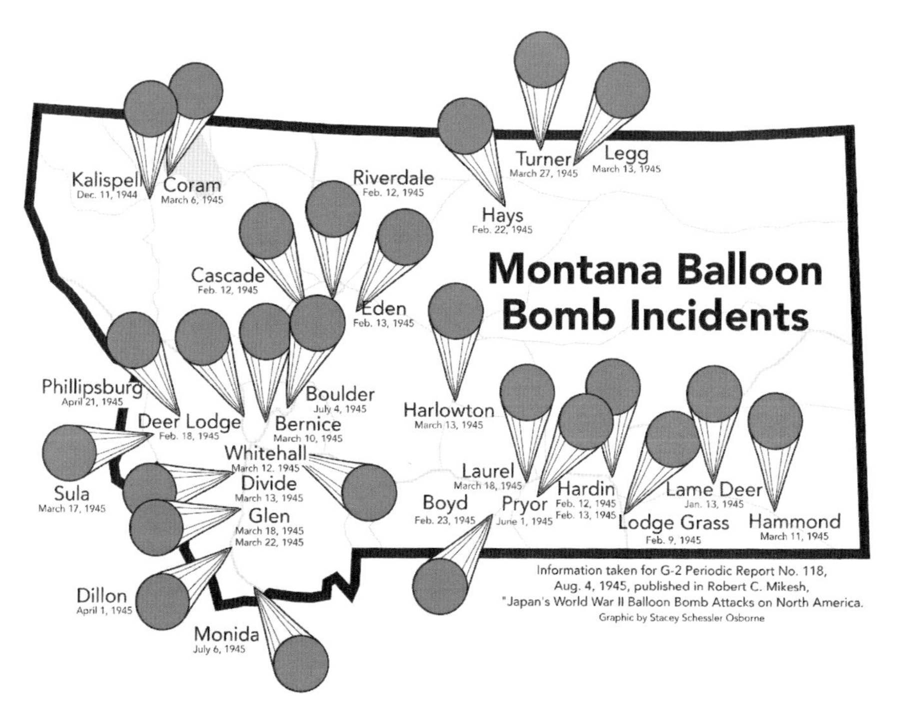

There were at least thirty balloon bomb discoveries in Montana in late 1944 and early 1945. *Stacey Osborne, Laurel Outlook.*

In an Associated Press interview after the end of the war, Japanese officials involved in the balloon program said they had been monitoring U.S. radio broadcasts as a way to measure the effectiveness of the bombs. The news blackout led them to conclude "that the weapon was worthless and [the] whole experiment useless."

While a significant number of the balloon bombs launched by the Japanese likely never reached North America, uncertainty remains as to whether there are more unexploded wind-driven weapons lingering in the woods near Flathead Lake or submerged in its waters. That possibility adds intrigue to what Michael Collins, director of the Smithsonian Air and Space Museum in the 1970s, described as "one of the most bizarre and obscure chapters of modern warfare."

PART III

The Science

CHAPTER 1

THE LAKE WATCHDOG

Morton J. Elrod, fresh into his University of Montana biology teaching career, took his first "collecting trip" to the Flathead Indian Reservation in 1897. A year later, with the idea of establishing a biological station near Flathead Lake, he reportedly explored every bay and cove along more than one hundred miles of the lake's shoreline in search of a suitable location. In 1899, Elrod chose a five-acre parcel on the north bank of the Swan River, about half a mile from where the river joins Flathead Lake.

The early mission at the Flathead Lake Biological Station—essentially a small building and some tents situated on land leased from Everit Sliter, the founder of Bigfork—was surveying and mapping the lake. The station, which offered classes and field trips in the summer months, also served as a base camp for Elrod and students to explore the surrounding mountains and streams. It was a remote, idyllic setting described by Elrod as sitting in a primeval forest at the foot of roaring rapids, with only two log houses in the area.

But within a few years, Elrod, concerned about growth in Bigfork and propelled by the fact that Sliter wanted to use his land for other purposes, was working to find a new location for the biological station. He got a big boost from Congress, which in 1905 agreed to provide 160 acres, carved from the Flathead Indian Reservation, to the biological station. Elrod selected forty acres on what is now Big Bull Island (formerly called Idlewild), another forty-acre parcel on Wild Horse Island and eighty acres at Yellow Bay, where he

The Flathead Lake Biological Station got its start in Bigfork but moved to Yellow Bay (pictured) in 1909. Morton J. Elrod photo. *Archives and Special Collections, University of Montana.*

would locate the main station operation. Elrod proclaimed Yellow Bay, with a protected harbor and a source of spring water, to be "one of the most charming spots on the shore of Flathead Lake." (The parcel on Wild Horse Island was later exchanged for property on Polson Bay.)

The station began operation in 1909 at Yellow Bay, and within a year, the Montana legislature agreed to spend $5,000 for the construction of several buildings and purchase of needed equipment. But even with the donated land and state funding, Elrod conducted a long battle to justify the station's existence to some university officials and to find additional money to keep it financially afloat. In a report to University of Montana president Clyde Duniway, Elrod noted that station had lived hand-to-mouth from its beginning and shared his sense of urgency about developing it. He wrote: "The country around Flathead Lake is fast settling up, the Reservation will soon be occupied, in a few years the wilderness will be changed to a rich and prosperous valley full of prosperous people."

More than 125 years after its founding, the biological station has a long-established national reputation for freshwater research and being a vigilant

Flathead Lake watchdog. It has weathered lean years during wars and the Great Depression and has been in the vanguard of numerous battles aiming to protect the lake and the vast ecosystem that nearly surrounds it from environmental degradation.

From a picturesque bay and thousands of feet of shoreline to the cluster of buildings on the thickly treed grounds, the station is a gorgeous scientific outpost. But its "laboratory" stretches far beyond Yellow Bay to the highest peaks of Glacier National Park and deep into the forests shrouding the three forks of the Flathead River that feed cold, clean water into the lake. But station researchers have gone much farther afield, having worked on almost every continent, gathering knowledge that can be readily applied in Northwest Montana.

"The bottom line is that the Flathead is still one of the most pristine, intact ecosystems in the populated world," said Jack Stanford, who worked at the biological station for thirty-six years and was its longest serving director before retiring in 2015. Credit for the success in the Flathead, he told this author, belongs to the isolated, focused atmosphere at Yellow Bay. "It's one of the most successful field stations in the world."

The *Missoula* at Yellow Bay, described by Morton J. Elrod as one of the most scenic spots on Flathead Lake. Elrod photo. *Archives and Special Collections, University of Montana.*

While some of the station's research may take place in beautiful locations, a critical aspect of the scientific work there involves a seemingly mundane activity: recordkeeping. Monitoring work is at the heart of much of the research on the lake and elsewhere in the Flathead ecosystem. The intensive lake monitoring began nearly forty years ago.

"There are probably only three or four places in the U.S. that have monitoring of this longevity," Jim Elser, who became the station's director in 2016, told this author. Such monitoring is the key to understanding biological change, whether it involves tiny freshwater shrimp, nutrients that can affect water quality, the effects of climate change or any of the other forces at play in Northwest Montana.

On a dark early July night in 2016, the author joined Elser and a couple of other researchers aboard the *Jessie B*, the biological station's research boat, as it left Yellow Bay and headed north to a spot near a buoy off Woods Bay. It was the first stop on a Flathead Lake sampling trip.

Members of the crew heaved a large cone-shaped net anchored by a heavy weight off the stern of the boat, and a winch lowered it to the bottom of the lake. It was retrieved quickly, gathering what it could on the trip back to the surface. Tiny creatures harbored in a plastic container at the bottom of the net were transferred quickly to sample jars.

This sampling is done monthly, and on each trip, the boat's GPS and sonar equipment guide the crew to the same points in the lake. The sampling at this first, northernmost stop is targeted to begin one hour and forty minutes after sunset, as close to the new moon as possible. A second stop that night was at a point closer to Yellow Bay, in some of the deepest water in the lake.

Of particular interest on these middle-of-the-night excursions are Mysis shrimp, tiny creatures that have caused a large-scale biological disruption in Flathead Lake. The Mysis hang out near the bottom of the lake during the day, rising up at night to eat zooplankton and other aquatic treats near the surface.

The shrimp were introduced into lakes upstream of Flathead in the 1960s and 1970s in an attempt to bolster a popular Kokanee salmon fishery. But the shrimp made it to the big lake in the 1980s, possibly migrating from nearby Swan Lake, and things went south in a hurry. The shrimp feasted on zooplankton, and their population exploded. Kokanee, which also relied on the zooplankton, were largely unable to feed on the bottom-dwelling shrimp. But lake trout in the deep water gorged on the shrimp, and their numbers skyrocketed. The voracious lake trout ate Kokanee, along with native cutthroat and bull trout. Scientists call this chain of events a trophic cascade, one that is likely irreversible.

Out on the lake, the Mysis netting is a key part of the efforts to monitor population trends and create a long-term record. Back at the biological station, the shrimp are counted and measured, their gender noted.

While the disappearance of the tasty Kokanee is old news, the Mysis continue to present a threat to the lake. Along with ravaging the native fish population, the tiny shrimp has also significantly increased the amount of carbon in the lake—a threat to Flathead's water clarity, one of its top attributes. "The arrival of the Mysis changed everything in the lake," Tom Bansak, the station's associate director, told this author in a 2023 interview.

In the mid-1980s, amid growing concerns about nutrient levels in the lake, Stanford and others at the station were key backers of a local option ban on detergents containing phosphorous. Despite stout industry opposition, local voters approved the ban. But concerns persist decades later as growth in the Flathead Valley and nearby areas has raised concerns about fertilizer and sewage and septic systems funneling damaging nutrients into the lake, which can spur algae growth and reduce water clarity and quality.

While there have been victories in the fight to protect the lake and its sprawling watershed that stretches across international borders, the list

The station's watershed monitoring work extends up the forks of the Flathead River, including the Nyack Flats east of West Glacier. *Author photo.*

of threats remains lengthy. Blinking red on the radar screen are aquatic invaders, including zebra and quagga mussels, which first hit North America in 1988 and have spread dramatically. The mussels suck up much of the food used by fish, and explosions of their numbers can leave razor-sharp shells scattered across beaches. The nasty creatures attach themselves to boats, trailers and fishing gear, which allows them to spread quickly and widely. Eradicating mussels in lakes in the United States has proven nearly impossible. Faculty members at the biological station have developed a monitoring system using DNA that can detect mussels from the presence of a few cells in water samples.

Scientists and researchers based at Yellow Bay also study the effects of climate change in the watershed, taking their work to mountainsides in Glacier and far up the Flathead River, which feeds the lake and plays a large role in its water quality. Bugs, more specifically stoneflies, are a key focus of the upstream work.

Meltwater and glacier stoneflies are two tiny species that are found only in Glacier. Reliant on water from glaciers, snowfields and high-altitude springs to survive, the stoneflies are imperiled by the climate-related shrinking of the park's glaciers. In the eyes of researchers, the stoneflies are an indicator species, an environmental bellwether.

Since the 1980s, the biological station has maintained a string of twenty monitoring wells along the park's southern border, in an area along the Middle Fork of the Flathead River known as the Nyack Flats. The wells provide a glimpse into what lies beneath in the aquifer: a subterranean realm of water, gravel and stoneflies, some of which exist without light and little oxygen. "If they are present in an area, then you have a pretty healthy, intact floodplain ecosystem," explained Rachel Malison, a stream ecologist and assistant professor at the biological station, in a 2019 interview with this author.

As Malison and other researchers note, the broader Glacier area is not the only place where stoneflies are found. But with its natural river flows, surrounding wilderness and a floodplain largely unaffected by development, the Nyack, unlike spots in Alaska or in the lofty reaches of Glacier, is more easily studied and has a long history of research, boosting its scientific value.

Despite being largely surrounded by tall peaks and wild land, the Nyack Flats site is, at least relatively, not remote. While it's remarkably pristine, the flats and broader Middle Fork region are a place of peril, as hundreds of cars, trucks and motor homes zip through the valley on US Highway 2. Parallel to the highway are the rails of the BNSF Railway main line that crosses northern Montana. One some days, dozens of trains rumble along the Middle Fork ferrying all sorts of cargo, including crude oil, a fact that concerns researchers and others. If even a few of those oil-hauling tanker cars were to tumble into the river, a spill could wreak far-reaching environmental havoc stretching to the lake and beyond.

While the forks of the Flathead River may represent a threat to the lake, they are also key contributors to the lake's clean, clear water. Fed by remarkably clean water from a large swath of pristine high country, the Flathead River essentially runs through the lake before exiting at Polson. Researchers say the flow has a flushing effect that allows the water in the lake to be replaced, on average, every 2.2 years. At Nevada's Lake Tahoe, which has less surface area but is much deeper than Flathead, the replacement time is more than six hundred years.

Keeping a watchful eye on water quality remains a large part of the biological station's mission, and that's been the case for years. Despite challenges and ever-changing threats, the story contained in the station's records, diligently maintained since 1977, is a happy one.

In 2023, Flathead Lake's water clarity was in the top 1 percent for temperate lakes in the world. Its phosphorous levels are among the lowest found in fresh water on the planet. "Flathead Lake is as clean today as it was in 1977," said Bansak, summarizing the state of the lake for guests at a 2023 open house, almost 125 years after Morton Elrod pounded the tent stakes near Bigfork. "We are a world-renowned water quality success story."

CHAPTER 2

THE RELENTLESS MORTON ELROD

When Morton J. Elrod left Illinois to come to Montana in 1897, the state had just 143,000 people, and Missoula, the home of a fledgling university, harbored roughly 4,800 residents. While Elrod's job official job was to teach and develop a science program, a biographer writing more than a century later contended the Midwest transplant had actually "succumbed to the allure of pristine wilderness."

In course of his forty-year career at the University of Montana, Elrod managed to leave many footprints across the peaks and valleys of western Montana and left lasting marks on the university campus, the city of Missoula and the fascinating region to the north, which includes Flathead Lake and what would become Glacier National Park and the National Bison Range. Elrod, a scientist at heart, was also an explorer, collector, methodical recordkeeper, prolific photographer and entrepreneur.

Elrod took his first collecting trip to the Flathead Indian Reservation in 1897. A year later, with the idea of establishing a biological station along Flathead Lake, Elrod soon chose a spot on the north bank of the Swan River near Bigfork.

The early mission at the station was surveying and mapping the lake. The station, which offered classes and field trips in the summer months, also served as a base camp for Elrod and his students to explore the surrounding mountains and streams. In 1902, Elrod published some of the findings that resulted from what he described as a "biological reconnaissance" in the Flathead Lake area.

While he roamed much of western Montana, Morton J. Elrod founded the biology program at the University of Montana. *Archives and Special Collections, University of Montana.*

Two years later, Elrod and a couple other men made their first trip to the pictograph site on lake's west shore that the scientist described as the "Pictured Rocks."

Elrod's explorations soon took him beyond the lake. His made his first trip to the area that would become Glacier National Park in 1907, a place he described as "fit for gods to dwell." As he did elsewhere, Elrod recorded many observations about the park's flora, fauna and glaciers. Some of his photographs from his early visits serve as the baseline of a project launched nearly one hundred years later to document the shrinking of the park's glaciers. Elrod was also the author of what most believe to be the park's first guidebook and helped found a naturalist program that led visitors on hikes and presented evening programs in the park's hotels. When the campaign to designate Glacier the nation's tenth national park gained steam, Elrod was an ardent supporter, joining George Bird Grinnell and others who persuaded President William H. Taft to sign the key legislation.

Along with his Glacier adventures, the first years of the twentieth century were filled with projects for the industrious Elrod. On behalf of the American Bison Society, he scouted potential locations for a bison range aimed at helping preserve the remnants of the huge herds that once roamed the western plains. One of Elrod's early ideas for the range was the roughly 2,100-acre Wild Horse Island. But a better location surfaced on the Flathead reservation near Ravalli, where the federal government paid $47,000 for more than 18,500 acres of land and the bison society pitched in $10,000 to buy thirty-four bison from area residents to establish the National Bison Range in 1909.

At roughly the same time, Elrod worked to find a new location for the biological station and landed on a location down the east shore at Yellow Bay. The station began operation there in 1909, and within a year, the Montana legislature agreed to spend $5,000 to allow the station to construct several buildings and buy equipment. But even with the donated

Above: Morton J. Elrod (*left*), a dog and an unidentified man on Wild Horse Island, a site of the scientist's early Flathead Lake study. *Archives and Special Collections, University of Montana.*

Opposite: Along with scientific work, Elrod boosted his income with other ventures, including promotional work for the Northern Pacific Railway. *Denny Kellogg collection.*

land and state funding, Elrod conducted a long battle to justify the station's existence to some university officials and to find additional money to keep it financially afloat.

While the station did get partial university funding in the early days (and still does), Elrod personally conducted fundraising campaigns. His donor list included a high-profile name: W.A. Clark, the Butte "copper king" who, after Montana gained statehood, battled Marcus Daly in a campaign to select the site of its capital city (Clark's Helena bested Daly's Anaconda) and later found infamy after he bribed Montana legislators to win an appointment to the U.S. Senate. The first donation to the biological station was $200, and

The Flathead Lake Country

NORTHERN PACIFIC RAILWAY

subsequent gifts continued for at least a decade, the motivation behind the gifts unclear.

Elrod's life and work presents a study in human complexity. While he was respected among students, alumni and colleagues and described in 1924 by the University of Montana yearbook as "the gray-haired, fun-loving doctor of bugs," others, including Elrod biographer and former UM president George Dennison, have pointed out Elrod's interest in and teaching of the discredited theories of eugenics.

Elrod worked for the first fourteen summers as the station's director for no pay. Desiring to supplement his professorial pay, the scientist engaged in a variety of entrepreneurial endeavors, including working as a private environmental consultant, selling seeds of native plants and trees and buying "villa" sites on Flathead Lake for $125 with plans to resell them at a profit. He also created postcards from his substantial photo collection and reportedly sold one thousand in a single summer.

Dr. Elrod Dies In Sleep At Age of 89

Morton John Elrod, A.B., A.M., M.S., Ph.D., a member of the State University faculty since 1897, two years after it was founded, died in

Dr. M. J. Elrod

Elrod suffered a severe stroke while working in his UM office in 1934. Unable to work, he died at home nearly two decades later. Missoulian, *January 19, 1953.*

While Elrod often expressed concern about the changes in western Montana brought by the arrival of more people, he also played a role in the population growth. Working for the Northern Pacific Railway, Elrod wrote and supplied photos for a sixteen-page pamphlet that promoted the impending opening of the Flathead reservation to homesteading by non-Native people and touted its potential for irrigated agriculture, orchards and livestock grazing. Elrod wrote enthusiastically of the river's canyon below the lake, which contained "a series of rapids of wonderous beauty" that elicit "admiration from all who see this place." He also noted the canyon's potential for the development of water power, claiming, "This river alone will afford electric power for many cities."

While Elrod had a clear desire to boost his income, he also had a reputation for generosity. Shortly after arriving in Missoula, he agreed to take over official weather observation duties and also maintain streamflow records for area rivers and creeks. Elrod donated his paltry government pay for that work to needy university students.

Elrod taught and mentored thousands of students, including some who led memorable lives: Jessie Bierman, who championed maternal and child health care across the United States and around the globe; Jeannette Rankin, the famed Montana pacifist who, while serving in Congress, voted against America's entry into both World Wars; and Harold Urey, a UM student and, later, an instructor whose work on isotopes led to a Nobel Prize in chemistry in 1934.

In 1934, while working in his office at UM, Elrod suffered a disabling stroke at age seventy-one that ended his career of teaching, research and collecting. He lived for almost two more decades, largely under the care of his daughter, before dying in his sleep at eighty-nine in 1953.

H.G. Merriam, himself a legend at UM, described Elrod as "perhaps the most active and valuable man to ever serve the university," noting that "he was a relentless man of great energy, alert intelligence and high vision." Elrod left behind extensive collections, painstakingly maintained records of all sorts and notable achievements in conservation. Thanks to his deep interest in photography, the university archives now include more than ten thousand Elrod images that document plants, animals, geologic features and scenes offering a remarkable portrait of Flathead Lake, Glacier National Park and many other areas of western Montana.

CHAPTER 3

A LOVE OF THE LAKE

Jessie Bierman lived and worked across the United States and in many places around the globe, studying and promoting improved health care for women and children and earning a reputation as a pioneer in the field.

Before her death at ninety-six, she carved a remarkable record of accomplishment: earning a bachelor's degree from the University of Montana in just three years, attending medical school at the University of Chicago's Rush Medical College, becoming one of Montana's first female physicians and later earning a master's degree in public health from Columbia University.

While Bierman worked initially in Montana, her career later took her to San Francisco, Washington, D.C., Hawaii, Germany, India and other places. But her heart seldom strayed far from home. Born in the Creston area east of Kalispell, Bierman developed an early love of the Flathead Valley in general and Flathead Lake in particular. Even while working, she found time to return to her summer home on the lake near Goose Island, along the west shore.

"I would rather be here than any place in the world," Bierman told Janelle Leader Lamb in an interview for UM's alumni magazine, the *Montanan*. "I've always had this same feeling, that this is where I belong."

Bierman's father, Henry, was an early-day freighter, operator of a ranch in the Big Draw area west of Big Arm and later the owner of a butcher shop in Kalispell. After graduating from Flathead High School, Bierman attended

Born in the Flathead Valley, Jessie Bierman lived across the United States and the world but always considered Montana and Flathead Lake her home. *Flathead Lake Biological Station.*

UM and finished coursework for her biology degree by spending a summer at the Flathead Lake Biological Station in 1921, mentored by Morton J. Elrod, who had founded the station about two decades earlier. "I was really tutored by this great man—for weeks on end—using a microscope for the first time, and doing all these thrilling things," she told Leader Lamb. "It was a wonderful experience."

Her summer at Yellow Bay was the start of a lifelong relationship with the biological station, the scientific work undertaken there and the people who did it. While some longtime residents recall Bierman conducting fundraising drives for the station, soliciting money from residents around the lake, she also didn't hesitate to pull out her own checkbook. In the early 1970s, she gave money to help match a grant that allowed the construction of housing for graduate students at the station. Just a bit later, when the station's research boat, a surplus Navy craft named the *Taluga*, needed a new engine, Bierman paid for the $10,000 job.

The *Jessie B* research boat, a key part of the station's lake monitoring, was named for the scientist who died in 1996. *Flathead Lake Biological Station.*

In a 1982 interview, she explained her generosity: "We think we know a lot about the natural processes of a great lake like this, there is a lot more we don't know. Most people believe the lake is so big, and flow through it so great, that it can't be polluted." She described the lake as a treasure for the nation. "We don't have any lakes like this one left. Taking the lake for granted is one of the biggest mistakes we can make."

When Bierman died in 1996, an obituary in a San Francisco newspaper credited her with being "a leader in the development of modern standards of early childcare and maternal health practices." The summary of her scientific and humanitarian achievements made scant mention of her affinity for Flathead Lake.

"She left her entire estate to the biological station when she passed away," noted Tom Bansak, the station's associate director, in a 2023 interview. "It included her lakefront home and cash." At the time, the donation was the largest the station had received from an individual.

The money created an endowed professorship at the station and also supports Bierman Scholar graduate students and visits to the station by internationally known scientists, helps pay for publications and provides general support of the work at the station.

Before her death, Bierman poured a bottle of bubbly across the bow of the station's twenty-eight-foot research boat, the *Jessie B*, a vital cog in the lake research and monitoring efforts. Jack Stanford, the station's longtime director who retired in 2016, offered this blunt assessment of Bierman's contributions: "Without Jessie, the biological station would not be what it is today."

Bierman offered this explanation for her passion for health care, Flathead Lake and the biological station to an interviewer about a decade before her death: "I don't rest until I have done something about something that needs changing."

CHAPTER 4

ON SHAKY GROUND

The front page of the *Daily Inter Lake* on April 1, 1952, along with the news of Joseph Stalin's views on the potential for World War III and the Midwest's presidential primaries, noted that the Flathead Valley had experienced a dinnertime earthquake the previous day. Despite the delivery of the news on April Fool's Day, the quake was no joke: the main shock was initially measured at magnitude 5.7 and was followed by a series of aftershocks that lasted nearly thirty minutes. The quake's epicenter was near Swan Lake, although the aftershocks were more widespread, with one centered near the middle of Flathead Lake.

"East Lake shore residents reported they felt a strong shake there, probably because they were near the center of the underground earth movement," the *Inter Lake* reported. Residents of that area described wood-frame homes creaking and pictures swinging on the walls. While damage wasn't widespread or severe, the two-story laboratory building at the University of Montana Biological Station at Yellow Bay was jolted and cracked enough to prompt its later removal. The structure, one of the first built at the station, was constructed in 1912.

While seismologists have since reanalyzed the quake and reduced its estimated magnitude to 5.2, it remains one of the most powerful quakes to hit the Flathead. The 1952 quake had tongues wagging, at least in the Polson area, where the *Flathead Courier* reported this observation from a lakeshore resident: "If my house slides into the lake one of these nights, please throw me a rope. If that doesn't work, throw flowers."

44TH

Moderate Quake Shakes Valley

Top: A 5.2 magnitude earthquake on April Fool's Day in 1952 shook the east shore of Flathead Lake and beyond. *Daily Inter Lake.*

Bottom: A cornerstone from the 1912 laboratory building at the Flathead Lake Biological Station. Damage from the 1952 quake led to the building's demolition. *Author photo.*

The Flathead Valley is clearly an active seismic area and has seen plenty of earthquakes over its recorded history. But much of western Montana is on similarly shaky ground. A series of seismic monitors detects as many as ten small earthquakes per day in the region. Many of the quakes are of less than magnitude 2.0, said Mike Stickney, director of the Earthquake

Studies Office, part of the Bureau of Mines and Geology in Butte, in a 2020 interview with the author. These small quakes are rarely felt by humans. About 3.0 magnitude "is when people start to feel an earthquake," he noted.

Certainly, none of the Flathead quakes measured over the last century have been of "California is going to fall into the ocean" intensity. The quakes in the Flathead in 1945 (5.3 magnitude) and 1952 (5.2) and a 5.0 magnitude shaker in 1975 centered near Creston are considered "moderate" by seismologists.

Stickney knows what lies beneath the Flathead as well as anybody. He completed a graduate school dissertation on seismicity in the valley and Northwest Montana in 1980. He and others have noted the presence of significant faults: cracks in the earth's crust where rocks slip and slide against one another. One, the Mission fault, runs along the base of the Mission Mountains from north of Missoula to Bigfork. Similarly, the Swan fault traces the western edge of the Swan Range. Other mapped faults include the Creston fault and the Kalispell fault.

Yet none of the strongest quakes in Northwest Montana can be clearly attributed to activity along these known faults. The explanation for the Flathead's biggest shakes appears to be a much larger web of faults known as the Intermountain Seismic Belt. While the seismic belt runs from southern

Mike Stickney, in a laboratory in Butte, has tracked seismic activity around Flathead Lake for decades. *Montana Standard.*

Nevada north through Utah and Idaho, the northernmost portion of this subterranean maze, which can be up to one hundred miles wide, stretches from Yellowstone National Park northwest to the Flathead Valley. "There are lots of small earthquakes along that zone and, infrequently, the larger ones," Stickney said.

While earthquakes are rare, Montana has experienced significant seismic events. A 1925 quake near Three Forks damaged buildings and opened cracks in the ground. In Helena, the fall of 1935 brought a terrifying series of quakes that killed four people and destroyed or damaged dozens of buildings. But the most tragic seismic event in state history occurred late in the evening on August 17, 1959, near West Yellowstone when a 7.5 magnitude quake brought the side of a mountain down along the Madison River, trapping campers and blocking the river to form what has become known as Quake Lake. At least twenty-eight people died as a result of the quake, most of them in the massive rockslide.

The historic 1959 shaking was felt across much of Montana and into Canada and eastern Washington. Near Polson, residents were awakened by the quake, which shook for more than a minute; dishes and other items toppled from shelves. The *Flathead Courier* in Polson said a patron at a drive-in movie theater near town reported that the quake "felt like someone was jumping on his car bumper."

While the most intense Flathead-area earthquakes have come with a main shock and aftershocks, there have also been "swarms" of smaller quakes without a main shock. In the 1990s, there was a series of small quakes centered in the Kila area, while the late 1960s and early 1970s saw the Big Arm, Dayton and Proctor areas endure a swarm of quakes that toppled chimneys, cracked windows and caused a bell to fall from a rural schoolhouse. Several of the quakes in that swarm were greater than 4.0 magnitude, including a 4.7 magnitude shaker in April 1969 that caused damage all along the southern and southwest portion of the lake. That quake was followed by 21 aftershocks within a month and 325 aftershocks between May 1969 and December 1971.

In early 1975, the Flathead Lake area was rocked again. A 3.8 magnitude quake near Somers early on January 31 was quickly followed by a 5.0 quake centered just a couple of miles to the northeast on February 3.

A looming question: Are Flathead Lake or Northwest Montana primed for a big earth-shaking event? A seismic hazard map produced by the U.S. Geological Survey highlights two areas of Montana at the greatest risk of a major quake. One is in southwest Montana, in the general area of the deadly 1959 Madison quake. The other is centered on the Flathead and Mission Valleys. Stickney noted a federal study that found significant earth movement thousands of years ago along the Mission fault in the Pablo–St. Ignatius area. The study showed at an earthquake about 7,600 years ago in that area may have had a magnitude of 7.5. "The fact that they have happened before indicates that they are likely to happen again in the future," Stickney said, noting that Montana saw quakes of 6.0 or greater in every decade between 1920 and 1960 but hasn't experienced one since.

There is no uncertainty about the unpredictability of earthquakes. Dr. Francis Thomson, who was the head of the Montana Bureau of Minerals and Geology, told news reporters that fact in 1945, after a significant quake that shook the Flathead and elsewhere. "It cannot be too strongly emphasized," he said, "that no one knoweth the day or hour at which future shocks might occur, and that anyone who engages in such precise predictions must be written down as either a fool or charlatan."

PART IV

The Islands

CHAPTER 1

THE WILD STORY OF WILD HORSE ISLAND

At more than 2,100 acres, with mountains, tall trees, wildflowers, grassy meadows and miles of shoreline bounded by a large, stunningly clean lake, a band of untamed horses and record-book bighorn sheep, the allure of Wild Horse Island is undeniable.

The big island also harbors countless stories, some rooted in centuries-old Native legend, others latter-day tales of optimistic homesteaders, eccentric dreamers, ambitious developers and early deaths. Remarkably, one of the most noteworthy stories features a happy ending. While two neighboring Flathead Lake islands hold mega-mansions and "No trespassing" signs, Wild Horse is largely publicly owned and undeveloped and one of the largest of Montana's state parks, visited by more than twenty thousand people each year.

Polson attorney John Mercer grew up on Rocky Point, northwest of town, and the family home had a clear view of Wild Horse. Hiking trips on the island and evening boat rides along its shoreline are cherished childhood memories. As a Montana legislator in the 1990s, Mercer helped lead a bipartisan measure to limit development of the state park and preserve its primitive state.

"It's just a special place," said Mercer in an interview in 2021. He has climbed with each of his children to the island's highest point. "It's like an old friend." Like many familiar with the island's history, Mercer marvels at the fact that much of Wild Horse, which has some private lots and cabins, escaped the grasp of a deep-pocketed individual owner. But for much of the past century, that outcome was far from certain.

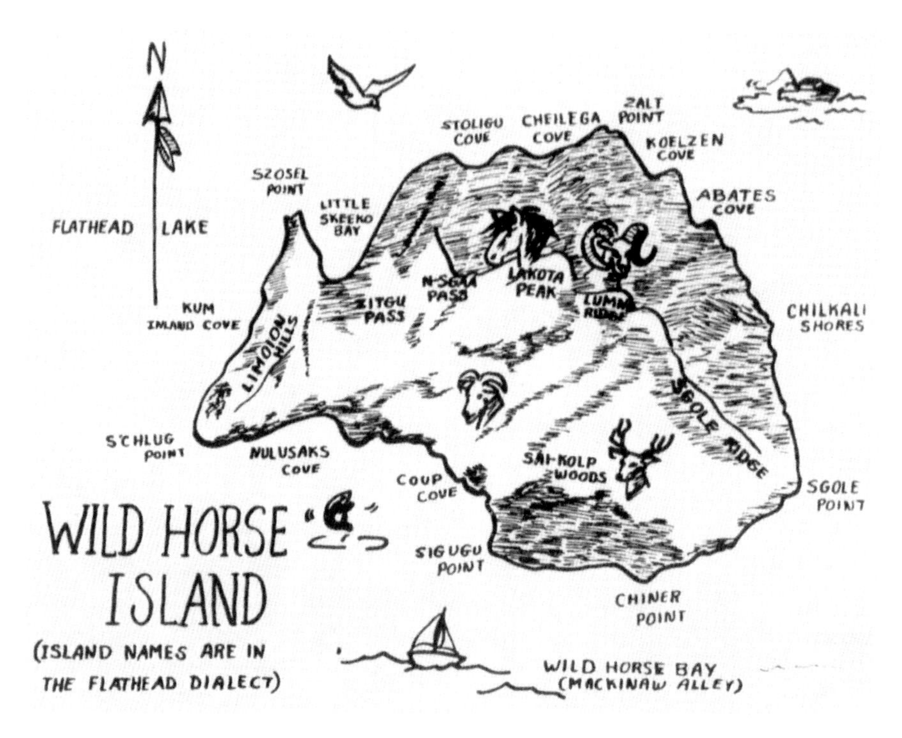

A map of Wild Horse from the 1960s featured Native names for points on the island. Artist unknown. *Missoulian.*

While details are few, the first "owners" of Wild Horse were the area's Native inhabitants, who referred to it as the Big Island. The island, along with the others in Flathead Lake, later became part of the Flathead Indian Reservation, an eventual product of the 1855 Hellgate Treaty between the U.S. government and the Salish, Pend d'Oreille and Kootenai tribes. Kootenai and Pend d'Oreille people lived along the lake's west shore.

While details are sketchy and at times conflicting, legend has the Native inhabitants using the island to protect horses from thievery by unfriendly tribes, including the Blackfeet. While the logistics of getting horses to and from the island, which is about a mile offshore, seem daunting, there appear to be elements of truth to the story. Kootenai historians have noted that horses and herders could have more easily reached Wild Horse by swimming a significantly shorter distance to and from what is now Cromwell Island, which, at least in periods of low water (and before the construction of Kerr/SKQ dam in the 1930s), was a peninsula connected to the lakeshore.

In his journals, White explorer John Mullan noted the presence of a band of horses on the island in 1854; he may be the source of the Wild Horse name. About fifty years later, the federal government offered allotments of reservation land to tribal members and opened the remaining parcels to non-Native homesteaders via lottery, a move understandably unpopular on the reservation. Some tribal members chose allotments on Wild Horse Island, joined later by White homesteaders with visions of raising crops and livestock in what seemed to be an idyllic setting.

But island life proved challenging. The W.A. Powers family planted fruit trees and a large black walnut tree on their claim, all of which were abandoned when they left the island shortly after the drowning death of their young son.

Another family, the Johnsons, completed a small house in December 1910, likely one of the first on island, and a nearby barn a short time later. The modest house was completed in just eleven days and may have been a community effort. It is unclear how long the Johnson family remained on Wild Horse. But 110 years later, the house and barn still stand along one of the island's main hiking trails.

McIntosh apple and pear trees believed to have been planted in 1918 are all that remain of another homestead closer to the center of the island. More than one hundred years later, some of the trees still bear fruit. There were short-lived attempts at running cattle and growing hay, but these efforts were likely stymied by a lack of water, an unanticipated challenge in a spot surrounded by a big lake.

In 1915, just five years after the island was opened to homesteading, many parcels on Wild Horse remained unclaimed. The federal government, rather than restoring the land to the reservation, put them up for sale. Of the 890 lots available, a man from Minnesota, Colonel Almond White, agreed to buy half, paying as little as fifteen dollars per acre for some of the parcels away from the lakeshore.

A real estate speculator and onetime associate of rail baron James J. Hill, White had grand plans: lakefront villas, a boys' school, summer camps for Scouts and Campfire Girls, a large hotel, a power plant on a creek and even a celestial observatory atop one of the island's high points. But the land sales didn't materialize, and White failed to make good on payments to the federal government. Many unsold parcels became available for anyone willing to pay back taxes.

The Reverend Robert Edington and his wife, Clara, who lived near Dayton, bought some parcels from White and later acquired forty additional acres on the east end of the island. The couple, who came to Montana from New York, worked to create a dude ranch that would cater to tourists from

The Hiawatha Lodge attracted guests from all over the country to Wild Horse Island. H. Schnitzmeyer photo. *Denny Kellogg collection.*

the East. The Edingtons constructed Hiawatha Lodge, cabins, stables and other structures and opened for business in 1931.

The couple entertained guests with boating and horseback rides from the lodge, which is likely the most significant structure built on Wild Horse. But in October 1934, a powerful afternoon storm hit the island, causing waves to wash over the lodge's boats. Edington and a caretaker rushed to the dock to secure the boats, but the waves knocked Edington into the water, and he drowned. His wife left the island soon thereafter and never returned.

Lewis Penwell, a Helena attorney, politician and sheep rancher, bought the lodge and the remainder of the Edington property and operated the dude ranch for a period before hatching a plan to turn the island into a game refuge. Penwell brought mule deer, turkeys and bighorn sheep to Wild Horse.

Whether the bighorns were the first on the island is unclear. Montana Fish, Wildlife & Parks (FWP) has photos of Penwell bringing bighorns to Wild Horse in 1939. Some accounts credit White, the land speculator, with introducing the sheep, while a publication from the Kootenai Cultural

Committee says tribal members captured bighorn lambs in British Columbia, nurtured them and later released them on Wild Horse, where they multiplied and started the herd.

"It is possible that there were other attempts to plant sheep," said Amy Grout in a 2021 interview. The Montana FWP ranger, who oversaw the lake islands managed by the agency, noted that she believed it is unlikely any early transplants survived.

Penwell, while stocking game animals on Wild Horse, also worked to buy all the land on the island, likely with an eye toward a future sale. While he purchased parcels from private owners, Penwell also worked in the political realm to transfer a thirty-six-acre parcel given to the University of Montana for biological and educational purposes to private hands. The original grant to UM also included parcels at Yellow Bay and on Bull Island.

Penwell did find a buyer for the island, a colorful New Jersey man named John C. Burnett, who had visions of an Arabian and Thoroughbred horse operation, all of it financed by his wife, Cora Timken, an heiress to a family roller-bearing fortune. While Cora visited the island only once, she took ownership in 1943. A few years later, she also gained title to the thirty-six island acres that had belonged to the Flathead Lake Biological Station, which, in an exchange, got eighty acres of less desirable land along the lake near Polson.

As part of the purchase, J.C. Burnett demanded that Penwell remove the island's horses. The horses, likely remnants of the dude ranch operation, were herded onto a barge that would ferry them from Wild Horse. But as one load made its way, the barge encountered strong wind and waves and tipped; the horses tethered to its deck were forced into the water. According to historical accounts, Penwell's son, Fred, jumped into the water and used a knife to cut the tethering lines and free the struggling horses, which swam back to the island, where they remained.

Burnett eventually built a herd of about one hundred horses on Wild Horse, using Arabian and Thoroughbred stallions to improve the herd. The New Jersey man reportedly enjoyed donning a black cowboy hat and roaming the island on a tall horse. In 1982, the late Missoula writer Bryan Di Salvatore described Burnett—who proclaimed the Montana sunshine he found on the island "to be the finest in the world"—as being "drunk with the whiskey of western fantasy."

Burnett's Wild Horse love affair didn't last. After several dry summers in the early 1950s left the island overgrazed and winter brought deep, crusty snow, Burnett had feed flown to the island for the horses. When Cora Timken

Burnett died in 1956, Burnett, despite inheriting 50 percent ownership of the island and an estimated $55 million, never returned to Wild Horse.

When Burnett died in 1959, a trust assumed ownership of the Montana island. Within months, Lake County commissioners, worried about a potential loss of property taxes, torpedoed an early proposal from conservation interests to buy the island and place it in public ownership.

Enter Bourke MacDonald, the owner of a multigenerational Butte lumber operation. MacDonald bought the entire island in 1961 for $240,000, a move described by his daughter-in-law as a spur-of-the-moment decision. "I think he needed a change of pace and he heard about this island," said Paddy O'Connell MacDonald of Missoula in a 2021 interview, noting that Bourke initially spent many weekends on Wild Horse and, later, entire summers.

Like others, MacDonald had big plans, at least initially: a hotel, stables, lakefront lots and a marina. But he also recognized the unique nature of the island, its wildlife and history. MacDonald worked to clear lakefront lots, punched in a road and thinned many trees. He ended up selling about fifty lots on the island's west and south shores, all of them circular, each with lake frontage but spaced to allow public access to the island's interior between the boundaries.

Bourke MacDonald died in his sleep on Wild Horse Island in 1973. His sudden death spurred a family discussion about the future of the island. Paddy MacDonald said her spouse, the late Missoula attorney Ron MacDonald, first hatched the idea of donating the island to the state, a plan embraced by others in the family. Wild Horse, she said, "is a siren call for the MacDonalds. My husband just held it in sacred regard, as did his siblings."

The transfer deal, hammered out in the late 1970s, was complex and controversial, involving legislative approval and funding from state and federal agencies and conservation groups. The bottom line was that the MacDonald family agreed to sell much of the island at half of its appraised $3.5 million value and donate the other half to the state to allow it to be preserved and publicly owned. The private land on Wild Horse today is just a 50-acre sliver of the island's 2,160 acres.

These days, all that's left of the three-story Hiawatha Lodge is a stone fireplace. But the bighorn sheep, wildflowers, rich human history and wild horses remain, along with a unique feeling of solitude that the big island exudes. In 2016, the skull and horns of a bighorn ram discovered on Wild Horse were believed to represent a world record for the species. With a Boone and Crockett Club score of 216-3/8, the ram beat the existing record by nearly seven inches. Officials said the ram, which appeared to benefit from

Right: While the population size has varied over the decades, wild horses remain on the island. *Author photo.*

Below: The skull of a bighorn ram found on Wild Horse in 2016 is believed to be a world record. *Author photo.*

the efforts to maintain the island's bighorn habitat, appeared to be about nine years old when he died, likely of natural causes. But managing wildlife on Wild Horse can be challenging. In 2022, game managers euthanized three mountain lions on the island after they became habituated to people and were suspected of preying on the bighorn sheep herd. FWP officials speculated that the lions, like the occasional bears spotted there, either swam to the island or crossed on winter ice.

Noting the mix of Native and White culture, the wildlife and wild tales, Grout, the FWP ranger, marveled at the foresight of the MacDonald family and many others from around the state who had the vision to preserve it. "To me, the island brings so many things together in one place," she said. "We are just blessed to have it as a public park."

CHAPTER 2

THE ALMOST-FORGOTTEN LIFE OF HERMAN SCHNITZMEYER

Herman Schnitzmeyer was lured west by a dream and the promise of a good life on almost-free land on a large island in Flathead Lake.

Schnitzmeyer was also lucky, at least in his early years. Born in 1879 in Illinois, Schnitzmeyer, a photographer and the son of German immigrant parents, was one of more than eighty-one thousand people who entered a lottery in hope of winning the right to select a homestead on the Flathead Indian Reservation. In about 1910, Schnitzmeyer filed a claim to 160 acres on the southeast portion of Wild Horse Island, with ideas of an idyllic life filling his head.

While Schnitzmeyer planted crops and built a dwelling on the island, the reality of an isolated, agrarian life didn't seem to suit him. "He wasn't much of a farmer or rancher, more of a philosopher," noted Denny Kellogg, a local historian. While on the island, he kept journals filled with his musings about life and photography. Schnitzmeyer was a prolific writer and schooled philosopher, "mostly for his own consumption," Kellogg told this author in 2023.

While Schnitzmeyer did prove up on his homestead claim, likely the first person to do so, island life wore thin. He spent the winter of 1913–14 on Wild Horse, likely its only human occupant. He battled isolation and, by his own account, starvation. A self-portrait made shortly after that winter shows a gaunt, wild-haired Schnitzmeyer, a possible reflection of his harrowing experience.

Left: Denny Kellogg, a Bigfork area historian, holds a self-portrait of Herman Schnitzmeyer made shortly after a lonely, hungry winter on Wild Horse. *Author photo.*

Opposite: Schnitzmeyer photographed a number of tribal leaders, including Kootenai chief Koostahtah. *Denny Kellogg collection.*

"His Wild Horse Island days were hard on him—physically as well as mentally," Polson newspaper editor Paul Fugleberg wrote nearly a century later. While on the island, Schnitzmeyer developed a photo postcard business with Louis Desch, another area homesteader. He also hatched a plan for a photo studio in Polson. In a 1914 journal entry, he outlined this vision: "My ambition now is to make a nationwide reputation for depicting the sentimental beauty of natural scenery, and while doing this, accumulate a nice competency."

Schnitzmeyer was a skilled photographer, especially in an era of bulky cameras and glass-plate negatives. He doggedly pursued images of Flathead Lake, the surrounding mountains, steamboats and Native inhabitants. He captured historic moments, including the first flight of a "hydroaeroplane" in Montana by aviator Terah T. Maroney in 1913 and the 1930 groundbreaking for what would become Kerr Dam.

He also captured the moments of everyday life in Polson, Fugleberg wrote, noting that "his camera clicked on sweaty threshing crews, lumberjacks,

ice-cutting operations on the lake, school activities, community plays, tribal events and chiefs, Boy Scouts, World War I doughboys, landmarks, buildings, people, hotel card games and parties. He even took pictures of corpses in caskets."

Schnitzmeyer opened his photo studio in 1912. A jovial sort, he also developed a reputation for unreliability, often missing appointments. "He was

Schnitzmeyer's photos often were of routine subjects, including this "cowboy band" in Polson in 1922. *Denny Kellogg collection.*

a very poor businessman because he would rather be out in the mountains," Kellogg said. Desch, his partner, had more business and marketing sense and helped produce hand-tinted images of Schnitzmeyer's work. The two produced a series of photo cards depicting scenes of Polson and the surrounding area that proved popular with homesteaders, who sent them to family elsewhere.

Schnitzmeyer also developed an appetite. Acquaintances noted his fondness for breakfast—he sometimes ate two in a morning. As a young man, Bill Gregg drove Schnitzmeyer around the area in a Ford Model T and recalled that on numerous occasions, the photographer would ask to stop at farmhouses and offer to trade a photo of the home for a meal. Odd behavior aside, "that man was a genius with a camera," Gregg told an interviewer.

Schnitzmeyer sold his photo studio in 1922 to Julius Meiers, an apprentice, who operated Lake City Studio in Polson for many years to come. Between 1922 and 1930, Schnitzmeyer did freelance photography work for the Northern Pacific Railway, producing images along the carrier's route in Montana and the Pacific Northwest that were used to promote passenger travel and the freight business.

In 1926, Schnitzmeyer, in need of money, sold some of his camera equipment and images to a man named Johan Rode, a Polson acquaintance, who duplicated some of the images, most of which were not copyrighted by Schnitzmeyer, and sold them as his own. "That's another reason he was not financially successful," Kellogg noted.

Long after Schnitzmeyer's death, Kellogg received a phone call from Roger Stang, Rode's grandson, who offered to give him Schnitzmeyer's Eastman View Camera No. 2 and related items. Kellogg accepted the offer without hesitation and drove to California to get the camera gear.

While some of Schnitzmeyer's photos were donated by the Desch family to the University of Montana decades after his death, other photos popped up in surprising places around Polson. Fugleberg, the newspaper editor, recalled getting phone calls from folks who found caches of the photographer's work; in one case, the photos were discovered under the floorboards of an old lumberyard.

After his Northern Pacific work ended, Schnitzmeyer landed in Missoula. While he did several talks about his photography, and UM hosted an exhibit in 1932 that included two hundred of his images, he faded into obscurity in the ensuing years, living in a hotel near the Milwaukee depot along the banks of the Clark Fork River. He also began to suffer physical and mental health issues. "He began to be somewhat paranoid," Fugleberg

View of early-day Polson, as captured by Schnitzmeyer. Date unknown. *Flathead Lake Biological Station collection.*

wrote years later, "and was obsessed with the idea that the British were trying to take over America. He went around insulting people he suspected of being in on the plot."

By the time of his death in 1939, Schnitzmeyer was destitute. He was buried at public expense in a Missoula cemetery with a simple grave marker, his name misspelled on his death certificate. Schnitzmeyer died of an abdominal obstruction, likely the result of a late-in-life diet of just milk and bread. The man who almost starved on Wild Horse Island weighed nearly three hundred pounds at his death. When his casket was being carried for burial, its handles broke off due to the weight. An amateur historian who lives near Bigfork, Kellogg has amassed a collection of eight hundred to nine hundred items related to the photographer, including many images, publications featuring his work, glass-plate negatives and the photographer's original gravestone.

"I am interested in artists whose legacy has been forgotten because no one was there to continue promoting them," Kellogg told this author in 2023. "Schnitzmeyer, unfortunately, became almost forgotten even before his death, perhaps because of his mental deterioration and no close relatives to support him."

Kellogg's collection of images and items related to the photographer are an attempt to memorialize the value of Schnitzmeyer's work and "his ability to capture a dynamic turning point in western Montana's culture and economy as it transcended from the Old West to the New West. His photographs were widely used to promote the expansion of white settlement in western Montana."

In the early 1990s, Fugleberg produced a booklet capturing the life and work of a man he described as a "homestead-era photographic artist." More than sixty years after the photographer's death, Kellogg located his grave marker, largely buried by dirt and grass, and worked with a Missoula monument business to place a new headstone on Schnitzmeyer's grave. The new stone includes a motto favored by the enigmatic photographer: "Love for Motive; Reason for Guide; Will for Strength."

CHAPTER 3

MASONS AND MERIT BADGES

The largest and best-known islands in Flathead Lake sit in the lake's Big Arm about a third of the way up the lake from Polson. Wild Horse Island, at roughly 2,100 acres, is the largest of the more than twenty named islands. To the northwest of the big island sits Cromwell Island, known to early residents as Papoose Island. At 350 acres, Cromwell is the second-largest island in the lake.

Sitting south of Wild Horse is the significantly smaller Melita Island. While only sixty-four acres, the island comes with a sizable story featuring likely use by Native inhabitants, long-term ownership by a somewhat secretive fraternal organization, decades of use by Boy Scouts and a mysterious, deep-pocketed benefactor.

Early in the twentieth century, much of the land on the Flathead Indian Reservation was split into parcels that were allotted to Native residents in an attempt to convince them to take up an agricultural lifestyle. After a few years, allotments unclaimed by tribal members were offered to homesteaders. In some cases, larger parcels of "surplus land" on the reservation were offered for sale to any and all.

In 1915, enticed by the seclusion and natural setting of the lake's Big Arm, a Masonic order bought what was then known as Wilgus Island from the U.S. government for $2,550. The plan was to create a "Masonic mecca" for recreation and get-togethers, according to a 1927 article in a magazine published by the fraternal order. "The natural beauty of this island, together

Knights Templar, in full regalia, gathered on Melita Island in 1937. *TAF Collection.*

with the advantages it offers for fishing and boating, make it an ideal meeting place and playground," the article stated. It was also a secluded place to practice the rites of Freemasonry, an organization with roots that stretch to the Middle Ages and the Crusades.

By 1928, the Masons had constructed a large lodge building on Melita (the island's name is a reference to the Mediterranean island of Malta). Described in a news account as "a cavernous medieval lodge," the structure, built for about $6,000, included a massive stone fireplace, possibly an architectural nod to the origins of Freemasonry, an organization that counted stonemasons, some of whom helped build ancient European cathedrals, among its early members. Along with the lodge, other buildings arose, and the island was served with electricity by cable buried in the lake bed. For a large 1927 conclave on the island, cars and people were ferried there by the Hodge Navigation Company. "Beautiful as to the setting, the island was equipped with every convenience," an article in the *Flathead Courier* noted. "There were tent houses for everyone who stayed overnight and electric lights all over the grounds." A gathering in 1928 saw two thousand to three thousand Masons and their families come to Melita.

In the years after World War II, the Masons struck a deal with the Montana Boy Scouts organization that allowed Scouts from across the state and region to use the island as a summer camp. For decades, the Boy Scouts earned merit badges in areas that included sailing, swimming and orienteering and took memorable field trips to nearby Wild Horse Island. The annual lease payment for this idyllic camping experience was reportedly just a dollar per year.

But in 1975, the after three decades of Melita Island summers, the Boy Scouts got a startling notice: the Masons had decided to sell. While the Scouts had the right of first refusal, they didn't have the $300,000 needed to buy the island with its two and a half miles of Flathead Lake shoreline.

Masonic Island (now Melita), with Wild Horse Island in the background. Herman Schnitzmeyer photo, date unknown. *Denny Kellogg collection.*

The Masons had little trouble finding a willing buyer. The purchaser presented a plan to sell lots for homes on the island. But sales never materialized as Lake County officials balked at allowing a large number of septic systems on the island. Back on the market, Melita caught the eye of a Nevada couple, Fred and Harriet Cox, who thought the island could be a great Montana home for themselves and their children. The couple bought Melita and seventeen acres on the nearby lakeshore in 1988.

Fred Cox later admitted that he and his wife were mistaken about the island's home potential. They decided to instead build on the more accessible lakeshore. "We never were going to develop it, but we concluded we didn't want to build there," Cox told an interviewer in 2003. "So we asked ourselves, what will we do with this island?"

A technology entrepreneur who grew up Texas, Cox was also a Boy Scout in his early years. In 1998, the couple had begun allowing Scouts to again use Melita, which had seen only trespassing visitors for a number of years. Not wanting to sell the island and see it developed, the couple approached the Scouts with a proposal: they would sell the island to the Scouts for $1.5 million, which amounted to about half of its appraised value. An agreement gave the Scouts until January 2005 to round up the money.

In late December 2004, with the deadline looming, the Scouts had raised only about $300,000. It appeared that Melita, known to generations as Scout Island, was slipping away, as it had in 1975. But salvation came via the post office. In the mail arrived a check for $1 million, the senders identified only as an "anonymous couple from western Montana." While they were still $250,000 short, the big donation jumpstarted fundraising, and within a few weeks, the Scouts had enough cash to buy Melita, with enough left over to make some improvements.

Page 4C • Friday, December 31, 2004 INDEPENDENT

$1M donation puts Flathead island within Boy Scouts' reach

An anonymous 2004 donation helped the Boy Scouts buy Melita Island. *Helena Independent Record.*

In 2023, eighteen years after the mystery donation, Camp Melita Island, wholly owned by the Montana Council of the Boy Scouts of America, offered more than forty classes and the opportunity to earn thirty-five merit badges to girls and boys.

PART V

The Mysteries

CHAPTER 1

THE MYSTERIOUS CREATURE

It was about nine o'clock on a Friday night when the 160-pound test nylon line on Leslie Griffith's large fishing reel began to sing the opening stanza in what would become a colorful, controversial chapter in the lore of Flathead Lake. In the days that followed that evening in late May 1955, Griffith would unspool a fish story reminiscent of a Hemingway novel.

Described as a "freelance fisherman," Griffith had been living near Dayton, on the lake's west shore, lured by hope of catching a large white sturgeon and claiming a reward offered by a group known as Big Fish Unlimited, an organization of Polson residents aiming to bring economically beneficial publicity to the community at the south end of Flathead Lake. A fish over nine feet would net a prize of $500, while a fish greater than fourteen feet long would yield a captor $1,000. Members of the group contended that sturgeon lived in Flathead, a belief fueled by multiple reported sightings, including one by J.F. "Fay" McAlear, a leader of Big Fish Unlimited, who spoke of seeing a large fish, possibly an eighteen-footer, in the lake in 1951.

Griffith, sixty-three, who had reportedly caught a three-hundred-pound white sturgeon in the Snake River in Idaho, took the reward bait. He spoke of fishing unsuccessfully for nearly three months before snagging a big fish near Cromwell Island in the lake's Big Arm that May evening. The fish, he said, pulled all six hundred feet of his line from the reel and towed Griffith, alone in his sixteen-foot cedar boat, around parts of the lake for five hours. At one point, the fisherman said, he used a gaff—a device with a hook attached to a handle or pole—to contain the fish. The gaff blow, which apparently missed vital organs, caused the big fish to thrash about violently, nearly causing the boat to capsize, Griffith told interviewers.

Later, the fisherman said, he used another gaff to subdue the fish and towed it initially to a dock in Big Arm and eventually to Dayton, where he located a telephone. His first call, in the minutes after dawn, was to McAlear, the Big Fish promoter, who quickly drove to Dayton and, with the help of Griffith and several others, loaded the fish into a truck and headed for Polson.

The events in May 1955 remain the only reported catch of a white sturgeon in Flathead Lake, although sightings of very large fish and even serpentlike animals stretch back centuries in Native legend and in the more recent stories of White inhabitants. The singular status of what became known as the "Griffith sturgeon" sparked fascination and skepticism that have endured for decades. Could a white sturgeon, the largest fish known to exist in North America, be the legendary "Flathead Lake Monster"?

While ancient Kootenai legend includes vague accounts of monsterlike activity in the lake, possibly the first more modern account to gain credence came in 1889 from James Kerr, the captain of the *U.S. Grant*, a steam-driven boat that ferried passengers and freight up and down the lake. Kerr reported seeing what he initially thought was a large log. As the boat got closer, the "log," estimated at more than twenty feet in length, began to move through the water. Along with Kerr, many of the more than one hundred passengers on the boat reportedly spotted the mysterious creature, which disappeared from the surface after shots were fired from a passenger's rifle.

In the years to come, sightings of a monster of varying shape and size came in waves. The tally of sightings grew over the years, many of them reported in the pages of the *Flathead Courier*, the weekly newspaper in Polson. Some accounts were sketchy, while others came from sources deemed credible.

The late Paul Fugleberg, the longtime editor of the *Courier* and author of a number of books about the history of the lake and its environs, chronicled many of the monster sightings. In a column in 2015, he summarized the nature of some reports: "Some of the sightings have been attributed to hyperactive imaginations, playful pranks, natural phenomena such as wave action, shadows, lighting effects, logs and a large number of animals, including bears, horses, deer, elk, dogs, a dead monkey, a loose circus seal and even an escaped buffalo."

Despite the cynical tendencies of news reporters, Fugleberg never dismissed the possible existence of a monster of some sort in the lake and

The 1956 sturgeon catch reported by fisherman Leslie Griffith sparked many newspaper headlines. Julius Meiers photo. *Missoulian.*

wrote on numerous occasions of the sturgeon hauled ashore by Griffith in 1955. And he also shared a letter he received from noted Montana author Dorothy Johnson in 1962, in which Johnson urged caution in dismissing sightings. "I don't think the monster should be done tongue in cheek," she wrote Fugleberg. "You have eyewitness accounts by people who were scared and didn't think it was funny. I remember hearing (about) something in Flathead Lake more than 40 years ago, so I don't give the Polson Chamber of Commerce credit for dreaming it up."

An avid reader of accounts of lake monster sightings was Laney Hanzel, who worked for Montana Fish, Wildlife and Parks as a Flathead Lake fisheries biologist from 1960 to 1993. Hanzel, who died in October 2022, compiled accounts and conducted interviews with folks who reported seeing large fish or other unusual animals in the lake. His obituary noted that Hanzel, through the years, became "the world-renowned Flathead Lake Monster expert."

Hilary Devlin, in her work for the Flathead Lakers, the conservation group, spent a memorable afternoon with Hanzel as he shared some of the accounts and stories linked to the alleged mysterious lake dweller. She also worked with Hanzel to create a map marking the general locations of more than one hundred sightings going back to 1889.

Hanzel didn't like the "monster" label. "He called it a creature," Devlin said in a 2022 interview, noting the map indicates sighting of either a large fish or a creature. "When Laney told stories and talked about it, you weren't sure if he believed it or not. He just presented it in a way that you could make up your own mind."

Hanzel said some accounts painted a picture of a creature that vaguely resembled descriptions of a sea serpent or even the famed Loch Ness Monster reputed to live in a Scottish lake. "Most, other than large fish sightings, describe it as long and eel-shaped, round, 30 to 40 feet in length with large black eyes," he told *Missoulian* reporter Vince Devlin in 2017. "They say it undulates through the water like a snake."

While Hanzel may have largely avoided sharing his opinion about a monster, or the possible presence of sturgeon in the lake, others of a scientific bent have expressed doubts. Despite extensive netting, electroshocking, small submarines that have ventured deep into the lake and other types of research, including time spent underwater in the forks of the Flathead River, no sign of sturgeon or other creatures has been found.

"The odds of something liked that going unnoticed are pretty slim," Brian Marotz, a now-retired FWP biologist, told an interviewer in 2003.

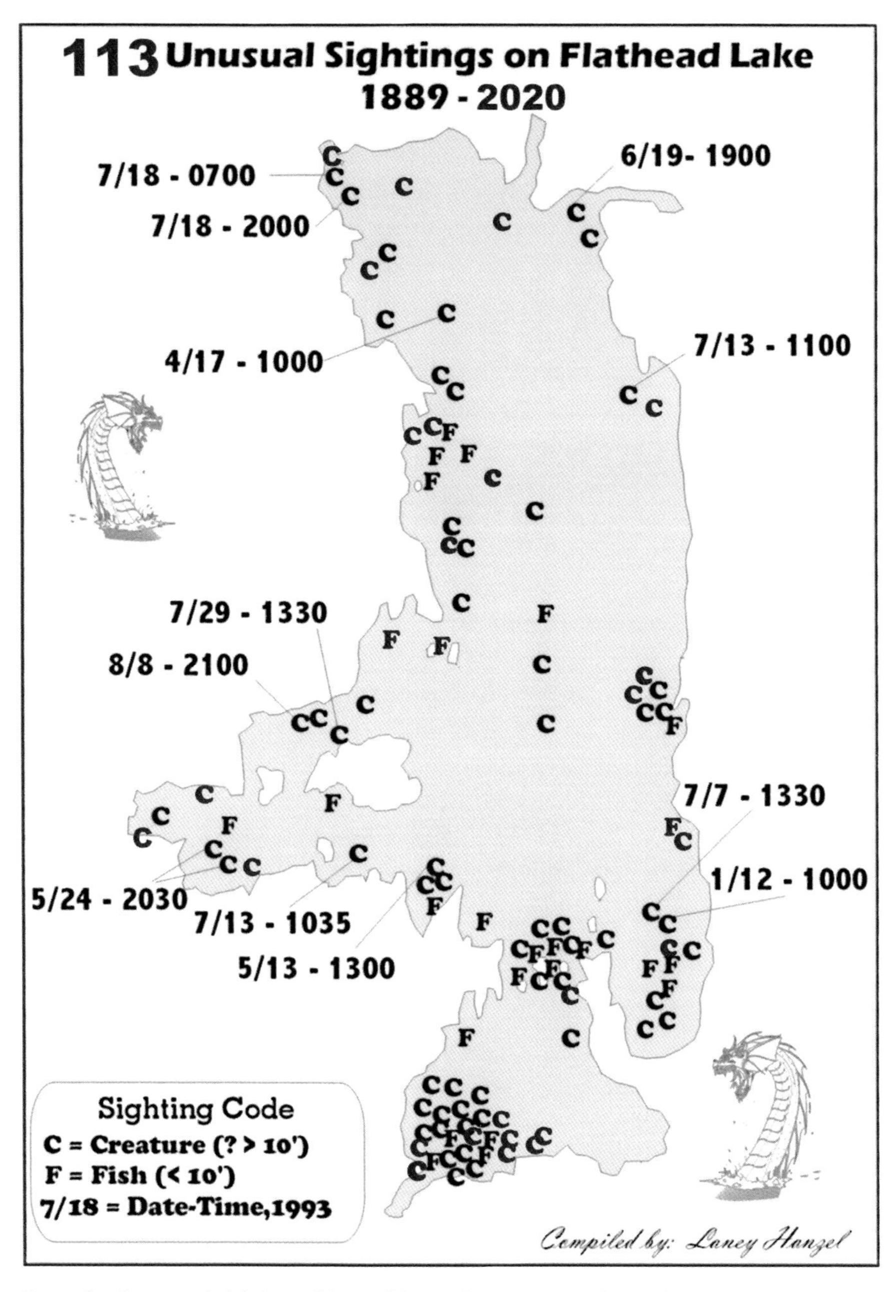

Records of reported sightings of large fish or other "creatures" were kept by a state fisheries biologist. *Map by Flathead Lakers.*

Given the popularity of fishing for lake trout in the lake's depths, he said, "you would think that someone would pick up a smaller sturgeon after all these years."

White sturgeon, the largest freshwater fish in North America, are native to several rivers in the Northwest that drain into the Pacific Ocean, including the Columbia system, which includes, in its upper reaches, the Clark Fork and Flathead Rivers. The big fish are found in Kootenay Lake in British Columbia, and there is a small, isolated population of white sturgeon in the Kootenai River in northern Idaho and in the river below Kootenai Falls, west of Libby. Biologists believe the presence of tall waterfalls in rivers and the construction of hydroelectric dams have limited the range of the white sturgeon in North America.

"There are lakes all over the Northwest that have sturgeon in them," noted Polson resident Karen Dunwell, who serves as president of the Polson-Flathead Lake Museum. As the granddaughter of J.F. McAlear, the Polson businessman and Big Fish promoter, her interest in sturgeon may be genetic. "I was bitten by the sturgeon bug," she said in 2022. "They are fascinating to me."

Dunwell reported that her grandparents spotted sturgeon while fishing off Finley Point decades ago. Years later, she said, she saw several sturgeon in the same area, "and they were easily more than ten feet." At another point, her husband spotted smaller sturgeon in the area.

Acknowledging the scientific skepticism about sturgeon or other creatures in the lake, she points to the fact that sturgeon can live for eighty to one hundred years, are elusive and can appear to be serpentlike due to their smooth, sinuous skin that allows them to move gracefully through the water. She harbors little doubt that the big fish live, or have lived, in Flathead Lake.

At the Flathead Lake Biological Station at Yellow Bay, scientists have studied the lake and its tributaries for more than a century. Tom Bansak, the station's assistant director, said there are no clear answers to questions about sturgeon, although the idea of a prehistoric creature seems far-fetched. In his view, the presence of white sturgeon in nearby waters and the oral histories of Indigenous people support the idea that the big fish could have lived in Flathead.

"I think it is highly likely that sturgeon have lived in Flathead Lake or its predecessors," Bansak said in 2023. "To me, the question is whether they still persist, and whether we can document proof of their current existence."

Within hours of its arrival in Polson, the Griffith sturgeon began to make waves. McAlear and others arranged to have the fish, more than seven

Posters created by Polson promoters declared the controversial sturgeon a "baby monster." *Author photo.*

feet long, placed on ice and put on display in the Salish Hotel building, where it remained for several days. Viewers who paid twenty-five cents (ten cents for children) were invited to guess its weight. McAlear shared colorful commentary, noting that the sturgeon "was just a small fish. The big ones are still out there....We think we've got the world's largest sturgeon out there."

Promotional posters described the sturgeon as "our baby monster" and "the biggest fish ever caught in Montana." Within a few days McAlear reported that at least seven thousand people had seen the sturgeon in Polson and at Allentown, between Ronan and St. Ignatius.

The apparent catch was front-page news across Montana and across the country. An account in the *Great Falls Tribune*, headlined "Fisherman Hooks Flathead Lake Monster," noted that the fish weighed 181 pounds, although "it subsequently swelled to 350 pounds in newsprint." The *Tribune* account also pointed out a critical fact: "The remarkable thing is no one has ever caught a sturgeon in Flathead Lake before, big or little, or produced any good evidence that sturgeon were there."

Some began to question whether the fish was actually caught in the lake. It was possible, some speculated, that the sturgeon was brought to the area from elsewhere as part of a promotional scheme involving Griffith and Big Fish Unlimited. In response, McAlear and Griffith signed notarized statements vowing the sturgeon was indeed caught in Flathead Lake. Promoters also came up with an additional reward, offering $1,000 to anyone who could prove the fish didn't come from the lake.

Two scientists associated with the biological station and the University of Montana in Missoula were unable to answer the question of the sturgeon's residency. After an examination of its innards the day it was brought ashore, zoologists Royal Brunson and Daniel Block said the viscera offered few clues about the origins of the twenty-seven-year-old sturgeon. But Brunson, who had done a ten-year study of Flathead fish, may have fanned the flames of doubt when he noted the presence of a fish backbone inside the sturgeon that didn't appear to come from a fish known to live in Flathead Lake.

In the months after the Griffith sturgeon surfaced, a number of residents and visitors reported spotting big fish in several parts of the lake, with some of the sightings reportedly involving a group of fish. Thain White, a well-known amateur historian and the operator of the Flathead Lake Lookout Museum south of Lakeside, reported watching a group of fish for several hours in September 1955. By his count, there were thirty-six sturgeon between three and twelve feet long in the group.

The sturgeon phenomena inspired additional anglers, including McAlear. Dunwell remembers spending many hours fishing for sturgeon with her grandfather, always plying the waters between Wild Horse and Cromwell Islands, where Griffith said he made his catch. They used wire cable for fishing line, a spool of which is on display at the Polson museum. She also recalled using animal blood supplied by a local butcher shop to chum the water.

The sturgeon story prompted James "Jack" Kehoe, who helped operate the *Helena*, the last of the steamboats to ply Flathead's waters, to share a sturgeon sighting account. In a letter to the *Daily Inter Lake*, Kehoe, who founded Kehoe's Agate Shop near Bigfork, recounted spotting what he believed to be a sturgeon in April 1927 off the mouth of the Flathead River. In the clear water, "a shadow about the size of a sawlog seemed to come up off the bottom right in our path," Kehoe wrote. At the helm of the boat, he cranked the wheel to avoid hitting the creature, which swam along with the boat for roughly one hundred feet before disappearing into deeper water.

"I could hardly believe my own eyes at what I had seen and while I remember every detail, I have never told anyone about it until people started to see the big fish lately, and then I was certain I hadn't been seeing pink elephants," Kehoe wrote.

Kehoe also objected to the bounty placed on sturgeon by the Polson promoters. Noting there was no record of anyone being harmed by the big fish, "why should we now want to start killing off our harmless natural resources?" he wrote. "I would be glad to contribute toward a prize for a good picture of one of these big fish free and alive but not a dead one, and I am sure a lot of other people feel the same about it."

While occasional sightings of a big fish or other creature were reported in coming years, the sturgeon mania appeared to cool, at least in the eyes of the public. But a simmering dispute between the fisherman and the Big Fish promoter eventually spilled into the legal arena. Griffith sued McAlear, claiming that he, not Big Fish Unlimited, was the rightful owner of the sturgeon, which had been mounted by a taxidermist, and that he was owed more money than the $698.48 he had received from its public display.

The case eventually landed in front of the Montana Supreme Court. More than four years after the sturgeon was trucked to Polson, in late 1959,

The Griffith sturgeon has found a home at the Polson-Flathead Lake Museum. *Author photo.*

the court confirmed that it belonged to Big Fish Unlimited, and while a more thorough accounting was needed, it ruled that Griffith was owed only half of the display proceeds. The Supreme Court justices did not offer an opinion, legal or otherwise, on the origins of the big fish or whether other sturgeon or mysterious creatures lived in Flathead Lake.

State fishing records kept by Montana FWP make no mention of the Griffith sturgeon. A ninety-six-pound white sturgeon caught by Herb Stout in the Kootenai River in 1968 is the official Montana record holder.

A mounted version of the controversial Griffith sturgeon, yellowed by the decades, hangs on the wall of the Polson museum, shrouded with questions about its colorful past. Dunwell said she enjoys the mystery that envelops the fish but doesn't see any answers to the questions about the sturgeon, or other possible creatures, surfacing soon. "You are going to have controversy about this forever," she said.

Chapter 2

COLD, DARK AND LONESOME

The Grumman F9F-8 was a fast, highly maneuverable fighter jet used in pilot training and by the U.S. Navy's Blue Angels to entertain thousands in the late 1950s. But how one of the jets, flown by a Montana-raised airline pilot on reserve duty, came to rest in the darkest depths of Flathead Lake in 1960 is an enduring mystery.

John Floyd Eaheart was raised in Missoula, played shortstop as a youth and was on a state champion basketball team before graduating from Missoula County High School. Attending what was then known as Montana State University (later renamed the University of Montana), he played baseball and basketball and received a degree in physical education. After graduation, he flew combat missions over Korea and remained in the military until 1957. He signed on as a pilot for Western Airlines in 1958 and served in the U.S. Marine Corps Reserve. (The Marine Corps is part of the U.S. Navy.)

In an interview in 2024, Cheryl Richmond, a Bigfork resident and niece of Eaheart, told of how her uncle played a father-like role in several lives, including hers. The young pilot was generous, kind and fun-loving. "He was extremely popular. Everybody loved Johnny. He would have given anybody the shirt off his back."

As part of his reserve duty, Eaheart completed training flights, a number of them to Montana. On March 21, 1960, he left Los Alamitos Naval Air Station in southern California, refueled his fighter at Hill Air Force Base in Utah and continued north to Malmstrom Air Force Base in Great Falls.

John Eaheart, no. 14, played on a state champion Missoula County High School basketball team in 1946. *Edie Opie Cope photo.*

After a short stay at Malmstrom, Eaheart filed a plan for a flight of about one and a half hours and headed west toward Missoula, making at least one pass over the city where his parents and a sister lived, at an altitude of less than one thousand feet, according to officials.

From Missoula, he flew north toward Flathead Lake, headed for the east lakeshore home of the parents of Vi Pinkerman, his girlfriend, who was in Denver, where she worked as a flight attendant. Eaheart, according to a Navy report, made several passes over the Blue Bay and Yellow Bay area as darkness gathered that evening, witnessed by several people, including K.C. Pinkerman, his girlfriend's father, who was outside his home. Pinkerman offered this account of the final pass to the *Missoulian* newspaper: "The plane came in from the south at an altitude of 600 to 700 feet and circled the Blue Bay and Yellow Bay area and then headed on a northwesterly course. At about 2,000 feet, it went into a left turn and about a 30-degree glide downwards from which it never came out."

Another witness, Mac Niccum, the owner of the Holiday Resort at Yellow Bay, reported seeing the jet buzzing the lake as he stood on the deck of his

The Daily Missoulian

87th Year. No. 327. Missoula, Montana, Tuesday Morning, March 22, 1960 5c

Jet Believed Down in Flathead Lake

News of the fighter plane that went into Flathead Lake on March 20, 1960, spread quickly. *Daily Missoulian.*

home. "It was going down and up and all around," Niccum told a reporter. "The last time we saw it, it headed off to the west pretty high. The next thing we saw was a water spout about 200 feet high."

Several witnesses, including Niccum, reported hearing an explosion after spotting the water spout. One witness reported hearing several explosions. Those reports were the source of considerable speculation in the days, months and years following the crash of the jet. It is unclear whether any of the witnesses saw the plane go into the lake.

Officials determined the plane entered the water at about 7:25 p.m., about forty-five minutes after sunset, at a spot roughly between Blue Bay and a point on Wild Horse Island. While reports of the likely crash sparked calls to law enforcement, searchers weren't able to reach the suspected crash site for several hours, and efforts to find Eaheart or the plane were hampered by darkness.

A key player in the search was the SS *Hodge*, a sixty-five-ton stern-wheeler work barge piloted by Frank Hodge, one of several generations of a family that traversed the lake in all variety of watercraft, including steamboats, over many decades. Hodge recalled the search effort to an interviewer in 1965: "They got me out at 9 that night and we covered the area all night with search lights. At 2 the next afternoon, we found some debris, about an acre of it, floating in the middle of the lake. We used a marked trolling line to find the depth. It was deep, 255 feet. All we could do was put out a marker buoy." Among the items found were a flight jacket, a pair of Navy dress trousers, a green battle jacket, a checkbook and a helmet that contained brain matter. Also found were wooden ribs from the plane's fuselage.

"A (Navy) salvage officer flew up from Frisco and looked the scene over," Hodge recalled, adding that the officer, referring to the jet, said, "'I'm going to recommend to the admiral that we leave it just where it is.'" Hodge added: "And that's just what we did."

Despite not having recovered the jet or the pilot's body, an investigative team from Whidbey Island Naval Air Station in Washington finished a

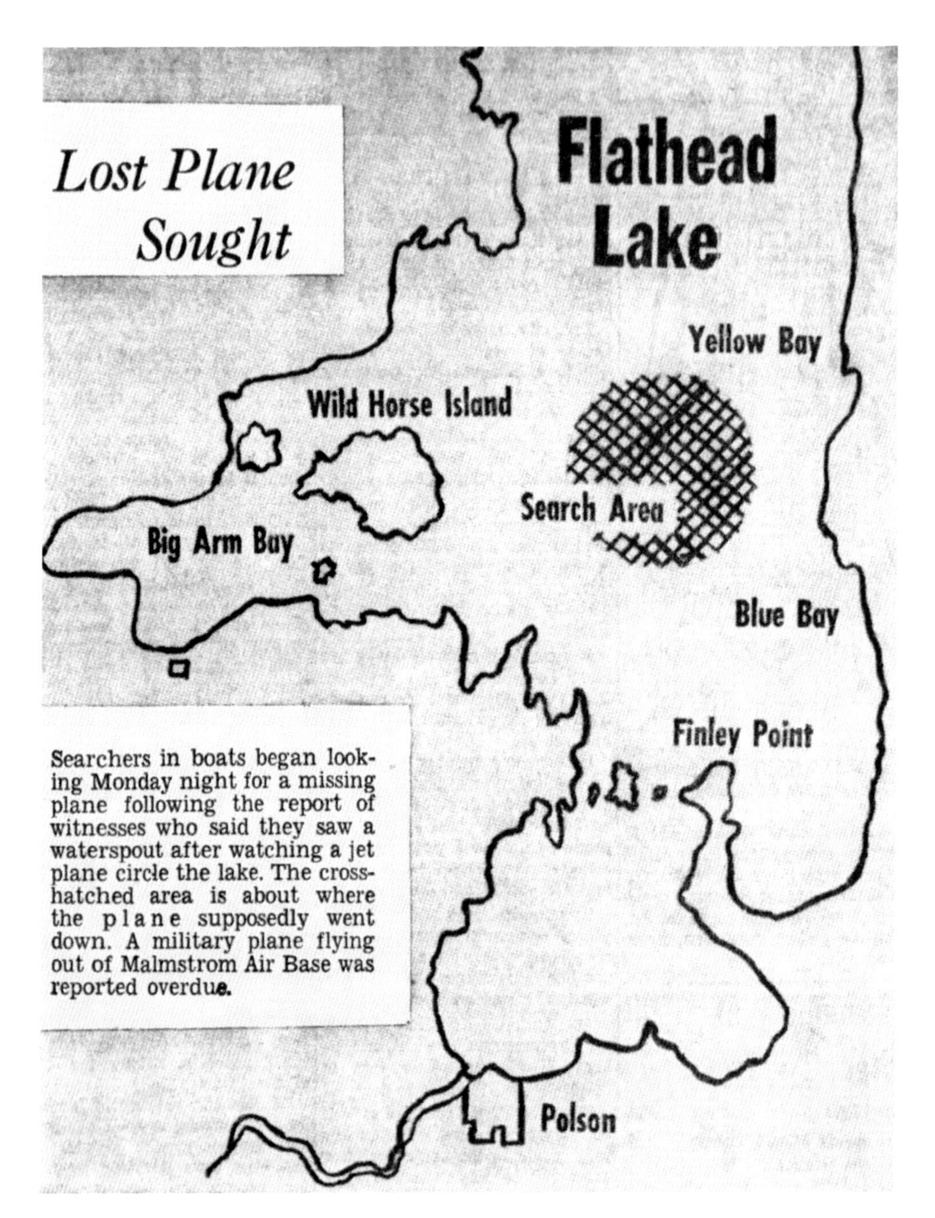

Lost Plane Sought

Searchers in boats began looking Monday night for a missing plane following the report of witnesses who said they saw a waterspout after watching a jet plane circle the lake. The cross-hatched area is about where the plane supposedly went down. A military plane flying out of Malmstrom Air Base was reported overdue.

The initial search took place in the deep waters between Wild Horse Island and Blue Bay. *Daily Missoulian.*

report on the crash in early April 1960, about three weeks after the crash. The report, which didn't find its way into the public realm until decades later, said there was no indication that structural or mechanical issues played a role in the crash. The plane, in service for thirty-nine months with 973 hours of flight time, had undergone a thorough inspection just six days before Eaheart's flight to Montana and had adequate fuel to continue flying. There was no distress call from the plane or evidence that the pilot tried to eject, which would have been proper procedure in a case of engine failure, according to the report.

The Navy report stated that the Eaheart's flights over Missoula and Flathead Lake were not part of the flight plan he filed at Malmstrom and stated that "it must be assumed" the pilot was aware that the low flights "were unauthorized and extremely dangerous." The report also noted that Eaheart had made a number of similar passes over the Pinkerman home on earlier training flights.

Noting the clear skies and calm conditions, the report speculated that Eaheart may have suffered from a loss of depth perception due to the glassy condition of the lake's surface and could have been looking toward the Pinkerman home instead of ahead at the immediate flight path when the plane hit the water. The location of the damage on Eaheart's helmet supported that theory, according to investigators. Other factors aside, the report said, "It is the opinion of the Board that the primary cause of this accident was unauthorized low flight."

Noting the recovery during the search of several ribs and pieces of fuel cells from the jet's fuselage, the investigators speculated that the impact of the jet hitting the water while flying at more than 350 knots may have caused the engine to smash through the fuselage and two fuel cells, causing the plane to break apart, possibly triggering the explosion reported by witnesses.

After the initial recovery of the helmet, clothing and fuselage components within about twenty-four hours of the crash, further searching in the following days yielded no more signs of the plane or Eaheart. On March 29, 1960, a Navy officer told Floyd Eaheart that there was no possibility his thirty-two-year-old son had survived the crash and the search had officially ended.

For nearly half a century, the story of the plane and the pilot lingered in the memories of family members, searchers and witnesses. A simple rectangular headstone memorializing Eaheart's birth and death was placed in the Missoula City Cemetery. Eaheart, who had been named to the

The Grumman F9F-8 Cougar was a fast, agile plane used by the U.S. Navy's Blue Angels flight demonstration team. *Pima Air and Space Museum.*

Grizzly Hall of Fame in 1955, was later honored for his basketball prowess with an annual award given to the UM basketball's top defensive player. Vi Pinkerman married Escoe Lewis, and the couple moved to Flathead's east shore, where they lived for decades.

But the story of the missing plane and pilot found new life in the spring of 2006. A private search took place, orchestrated by John Gisselbrecht, a Kalispell man, and involving the efforts of Sandy and Gene Ralston of Boise, Idaho, who donated the use of side-scan sonar equipment and a submarine-like remotely operated vehicle with a video camera. On May 2, Gisselbrecht said the searches spotted an aileron, a portion of a plane's wing, on the lake bottom at a depth of about 275 feet. A couple of days later, the searchers, joined by a cadaver dog—a black Labrador named Ruby who reportedly pawed at the water at the suspected crash site—located a Marine Corps boot, other remains and possibly a deployed parachute.

Gisselbrecht, who told reporters and others that he had been doing research into the crash for fifteen years, said he believed the crash was not the result of pilot error. Relying heavily on witness accounts of the crash, he offered a theory that the F9F-8 experienced issues that caused the engine to malfunction on its final pass over the lake. The pilot may have put the plane into a dive in an attempt to restart its engine. Gisselbrecht theorized that when that effort failed, Eaheart may have tried to eject from the plane moments before it went into the lake. He speculated that the plane's canopy may have failed to open properly and the pilot may have been critically injured when he collided with it.

"He was not hot-dogging or showing off," Gisselbrecht said at a press conference in Missoula in 2006. "He was in airplane prone to compressor stalls. This is a Marine who stayed at his post until the last possible second. It doesn't get any more honorable." He said his search effort, which didn't involve the Navy or local authorities, was focused on "restoring the honor of a Marine."

While they agreed that Eaheart was an honorable man, members of his family dispute much of Gisselbrecht's account, noting their opposition to the 2006 search. "He never ever found the plane," Richmond told this author. "The piece spotted in the lake bottom was not an aileron but a rudder from a boat," she said.

The 1960 crash occurred before aircraft were commonly fitted with flight data recorders; such a device could have offered information about the final flight of the fighter jet and possible engine problems. Even if the plane did have a flight recorder, known informally as a "black box," finding it on the silty lake bottom roughly 275 feet below the surface would be challenging.

Efforts to determine if the U.S Navy had done further investigation into the crash after the discoveries of 2006 were unsuccessful. A Freedom of Information Act request submitted to the Naval History and Heritage Command in Washington, D.C., in 2022 by this author was denied, with officials saying that, while the agency had documents and information related to the incident, that material was exempt from disclosure because it was "predecisional" and contained privileged information. "There's no further information available regarding this incident," officials wrote.

In an interview in 2019 as part of a Bigfork oral history project, Vi Lewis recalled the jet crash, relying largely on recollections of what she was told by her parents and others. She also offered another possible explanation for the crash: the jet piloted by Eaheart suffered mechanical problems after hitting a flock of geese. That possibility was the result of a conversation between her father and Frank Hodge, whose big boat was the first to reach the suspected crash site.

Lewis, who died late in 2022, watched the 2006 search activity from the shore and also went out on a boat on the lake as part of the effort. She told a reporter that she had mixed feelings about the possibility that Eaheart's body might be found. While she viewed the lake as too "cold, dark and lonesome" to be a final resting place, "I had learned to live with the fact that he was there and I didn't want him disturbed," she said.

Since the 2006 search, members of Eaheart's family have opposed any further efforts to search for his body or the plane. The death "was a terrible tragedy and great loss to our family," Richmond said. "As far as we are concerned, he went down in Flathead Lake and he is buried there." Another family member noted that there has been no effort by the Navy to find Eaheart's body or the plane. "I don't think they want to go get him. It's too deep and too cold." And in the view of the Navy, "it's ancient history."

CHAPTER 3
THE EXPLOSIVE MYSTERY OF MEDICINE ROCK

The chunks of rock that sit atop a tall ridge northwest of Rollins above Flathead Lake have long been the subject of mystery and speculation. In the eyes of the Kootenai, the fragmented chunks are pieces of a sacred rock they called Skinkuts, a reference to a coyote. To White settlers, it has long been known as Medicine Rock.

The rock sits atop a divide above the west shore. With the lake in the distance to the east, below is an area known as Big Lodge Flats, a three-hundred- to four-hundred-acre bench above the lake. Once the site of Native gatherings, the flats were the location for a stagecoach stop for several decades. The rock is very near an old foot trail, now overgrown but still discernable, used by travelers as a link between the upper Flathead Valley and the area at the foot of the lake and beyond.

The segment of the trail that passed near the rock was referred to by some as Medicine Rock Pass. Early settlers and some historians say the site has long been revered by the Kootenai and other early inhabitants and was, at least for a time, the location of large spring gathering, possibly in celebration of a victory in battle. In later decades, the rock was decorated by Native visitors with ornaments of various types, flags, animal skins and trinkets of many sorts.

In an interview in 1950, William Gingras, a Kootenai man, told area resident and historian Thain White that tribal members would place offerings at the site, often in a large crack on the top of the rock. It was, Gringras told White, "much like what you white people call a wishing well.

We Kutenais [Kootenai] used this rock the same way, just to wish for better times, good luck, more success, and the like."

A yellowed newspaper clipping tells of a pivotal moment in the more recent history of Medicine Rock. The headline: "Vandals Dynamite Historic Rock in Rollins District." The clipping relays the account of S.E. Johns, an early west shore homesteader who found the rock still standing but broken into large pieces. The "purpose of the dynamiters," according to the short article, "was evidently an attempt to find possible Indian relics."

White, in 1959, wrote that Fred Uhde, part of an early Rollins-area family, had shown him coins and other items that he said came from the rock. Among the loot were several five-cent pieces, the oldest one bearing a date of 1872; another was minted in 1880. There were some slightly newer Indian Head pennies in the collection. Some of the coins had holes drilled in them, as did a disk-like token emblazoned with "Vote for Bozeman for the Capitol, Nov. 8 1892." (Bozeman was one of seven cities that sought to be voted as Montana's capital that year. None of the cities won the needed majority, and Helena won the capital fight over Anaconda in a very contentious runoff election in 1894. The holes may indicate that the coins and token were once part of a necklace.)

Thain White reported that he searched the area near the rock with a metal detector and found beads, while others earlier had found Hudson's Bay blankets, moccasins and colorful cloth torn into ribbonlike strips. There

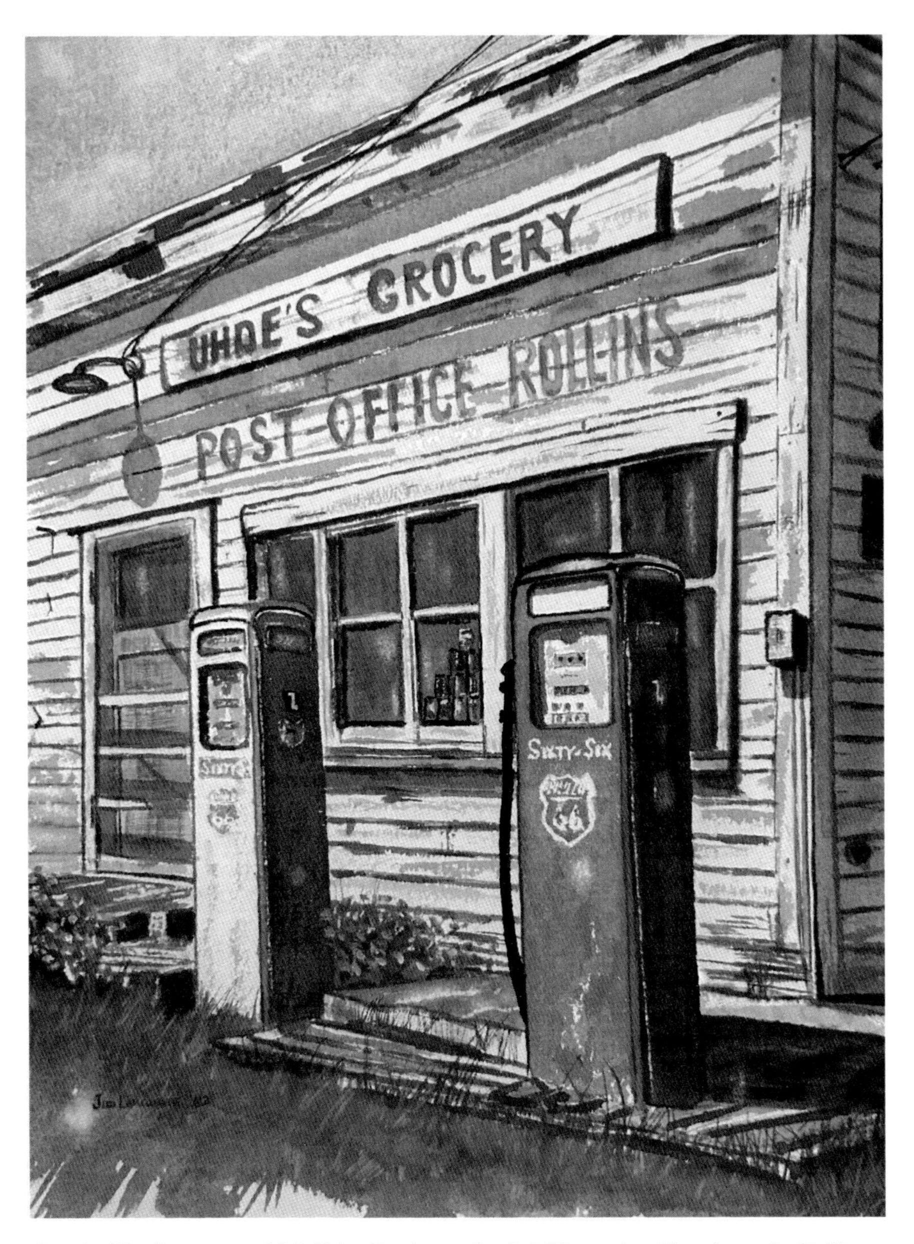

Opposite: The fragments of Medicine Rock remained visible on the ridge above the Rollins area in 2023. *Rick Weaver photo.*

Above: Uhde's Grocery, where some items from the dynamited rock may have been displayed. Painting by James R. Larcombe Jr., the author's father.

were also accounts of others extracting coins and other items from the crack in the rock before the dynamite incident.

Some believe that members of the Uhde family may have taken part in dynamiting the rock, which White said occurred in the late 1950s, but the news account of the incident appears to possibly be from 1937. In an interview in 2023, Rick Weaver, a retired Kalispell newspaper publisher and a grandson of Fred Uhde, said his grandfather and an uncle, Jim Uhde, could have been involved in the explosive incident.

"Both lived in Rollins at the time, and my grandfather ended up with several coins from the rock," Weaver said. "Plus, my uncle was a 'powder man' for the highway department. He knew dynamite and how much it would take to blow the rock apart."

The Uhde family came to the Rollins area early in the 1900s. For decades, Weaver's grandfather and grandmother operated a grocery store, gas station and post office along Highway 93. The store was also home to many artifacts and historical items collected by Fred Uhde. "When you went into the store, it was like going into a museum," Weaver recalled. As for the coins? Uhde may have given them to younger family members in the 1960s, and Weaver said the current whereabouts of the coins are unknown.

The property holding the Medicine Rock, owned at one point by the Burlington Northern Railroad, has changed ownership via corporate mergers and spinoffs over the decades. In 2021, it was sold to a Texas couple, who formed the 126,000-acre Flathead Ridge Ranch. In the summer of 2023, a group including Weaver; Eric Hanson, a Kalispell banker; and Bob Brown, a retired history teacher, legislator and state official from Whitefish, gained permission to visit the Medicine Rock and its environs. While the area has been altered by road construction and logging, after a short, steep climb, the group was able to find remnants of the old foot trail, resembling a narrow trough, which likely represents the first north–south route used by human travelers along the west shore of Flathead Lake.

They also claimed victory in finding nearby Skinkuts, the Medicine Rock, or at least the fragmented version of it. The rock, Hanson noted, sits on the only flat spot on the rugged ridge. "That was a huge rock," Weaver said. "It's got chunks that are squares that are as big as a desk." Referring to the mysterious explosion decades ago, he added: "It had to be done by someone who knew what they were doing."

Chapter 4

THE NIGHT THE MUSIC DIED

The Montana Band was riding high in the summer of 1987. The group's mix of country, bluegrass and rock was a hit in Nevada casinos and nightclubs, and the group had won further notice by being named the top country band at a recent Willie Nelson Music Invitational in Austin, Texas.

The band and its crew, including Flathead Valley resident Terry Robinson, boarded a ten-passenger propellor-driven airplane in Coeur d'Alene, Idaho, on July 4, 1987. They were bound for a show at the lakefront home of Dale Moore on Peaceful Bay near Lakeside. The twin-engine Beechcraft Model 18 made a low pass over the Moore home on its way to a landing at the Kalispell city airport.

After the show, band and crew boarded the plane and left Kalispell, headed back to Idaho for an evening performance in Post Falls. That flight also included a low pass over Flathead Lake and the Moore home shortly before eight o'clock. After flying near the home, witnesses said, the plane entered a steep climb, may have stalled and, after a sharp left turn, headed downhill at a steep angle.

"I could see the plane coming in over the house," Dellivan Thornton, who lived nearby, told a *Missoulian* reporter. "She was coming in real low and made a curve and then she just went down and hit between two trees." Another witness, Gene Holliday, reported seeing the plane come over the lake, start to climb and go into a roll. "I saw the angle and knew it was going in," he said.

The plane crashed through some trees, came to rest in an orchard and caught fire. Four of the plane's occupants were thrown from the aircraft, while six others remained in the burning wreckage, which took an hour to extinguish. There were no survivors in what officials would proclaim to be Montana's worst civil air crash.

Killed were Robinson; fellow musicians Kurt Bergeron, Cliff Tipton, Allan Larson and Grady Whitfield; the band's manager, Tom Sawan of Missoula; sound manager Dale Anderson, from Fort Benton; Tipton's twelve-year-old son Dallas Tipton; pilot Joe Taylor; and Taylor's friend Jean Lemery. News of the crash reverberated across the Flathead Valley, Montana and much of the music world. Newspapers across the country ran front-page stories about the crash.

In the minds of Montanans, the Montana Band was indelibly linked to its predecessor, the legendary Mission Mountain Wood Band, a group born in Missoula that attained a cultlike following for more than a decade. Mission Mountain was founded in the early 1970s by Robinson, Rob Quist and Steve Riddle, all onetime University of Montana students. Bergeron was also a member of the original band, and several others killed in the 1987 crash worked with Mission Mountain in nonperforming roles.

In the days after the crash, Quist spoke for many fans of both bands, telling an interviewer that those who died "carried the banner for the

Montana loses its top band

Feds probe plane crash that killed group

By LEN IWANSKI
Associated Press Writer

Investigators continued to sift through wreckage on a hill on the northwest edge of Flathead Lake today, seeking a cause for the plane crash that killed 10 people Saturday evening.

problem is that people aren't always sure what they've seen."

Paul, a second FAA investigator and one from the NTSB in Denver are analyzing wreckage of the 1946 twin-engine Beechcraft that crashed in an old apple orchard overlooking the lake, a popular resort area.

"There were so many eyewitnesses. The

The July 4, 1987 plane crash near Lakeside was believed at the time to be the deadliest civilian crash in state history. *Missoulian.*

Montana spirit, and that's the greatest gift of all. We felt we were really part of Montana and Montana was a part of us."

In Reno, Nevada, where the Montana Band performed regularly, news of the crash struck an emotional chord. Speaking to a reporter from the *Reno Gazette-Journal*, Paul Revere, a legend of rock, noted that he knew almost all the members of the Montana Band. "They were just the nicest bunch of guys," Revere said. "If they had the right break, they could have been as big as any country group in America. They were that talented."

A few days after the crash, more than 1,500 people attended a memorial service for the crash victims held at the Outlaw Inn in Kalispell. Quist and his Great Northern Band performed, and speakers remembered the victims. Jim Bergeron, who lost his son, shared words of consolation he had received from Montana governor Ted Schwinden, who told him, "All of us lost a bit of ourselves last weekend."

Local authorities speculated that up to two hundred people may have witnessed the plane come to the ground just a few hundred yards above Flathead Lake. The fact that the plane almost totally disintegrated in the fiery crash forced state and federal investigators to rely heavily on witness accounts to determine what happened on that holiday evening. There was public talk about engine issues or possible overloading of the plane. Others claimed the pilot was "showing off."

Bob Johnson, chief of the National Transportation Safety Board's Denver office, said the Beechcraft, built in 1945, was considered reliable. "That's a plane that has been around for a long time," he said a few days after the crash. "It's a workhorse, a very versatile plane that can carry a big load."

In its final report on the fatal accident, the NTSB concluded that there was no evidence of engine failure or other mechanical issues. The agency declared the probable cause of the crash to be related to the actions of Taylor, the thirty-eight-year-old pilot with about four thousand hours of flight experience. The pilot used poor judgment in flying too low and possibly attempting aerobatic maneuvers before the crash, the agency concluded, noting he may have also been overconfident in the plane's capabilities and his own flying ability.

Speaking to an interviewer in the wake of the Flathead tragedy, Steve Riddle, a Libby native, spoke wistfully of the Mission Mountain's high-flying performances and his former bandmates. "What we had with the Mission Mountain Wood Band was magic," Riddle said. "I'm not sure that magic will repeat itself again in my lifetime." He also offered another recollection from the early days: "We buzzed a lot of parties in a lot of airplanes over the years."

PART VI

Swimming the Lake

CHAPTER 1

THE ADVENTURES OF THE LAKESIDE FEMALE

In early September 2010, a young grizzly bear slipped into the water on Flathead Lake's west shore near Painted Rocks and headed south. Aided by webbed paws and a thick, hairy coat, she made it to Cedar Island on the first leg of what has become a legendary journey.

Known to biologists as the Lakeside Female, the four-year-old bear stayed a day on Cedar Island before paddling another three miles southwest to Wild Horse Island, where she hung out for three days. The next stop was the northwest corner of Polson Bay, where, after about a one-mile swim, she came ashore, shook the water off her fur and meandered in the foothills for several days.

On September 7, she headed east. The young grizzly swam first to Bird Island, took a little break and continued on the lake's east shore, a trip of about seven miles. From there, she scaled the Mission Mountains and seemed to find a spot to her liking near Swan Lake.

The story of the Lakeside Female's travels didn't come to light until nearly a year later. The bear had been fitted with a satellite collar in June 2010, not long before she started her swim. The bear, her mother and siblings first got the attention of bear researchers and others after they were captured near Bigfork, the prime suspects in a string of chicken killings. The Lakeside Female was relocated across the Swan Range to the west side of Hungry Horse Reservoir. But she didn't stay long and turned up next on Flathead's west shore. Whether her trip to that area came via a lake swim or a long hike is unclear. But her whereabouts from that point were carefully documented.

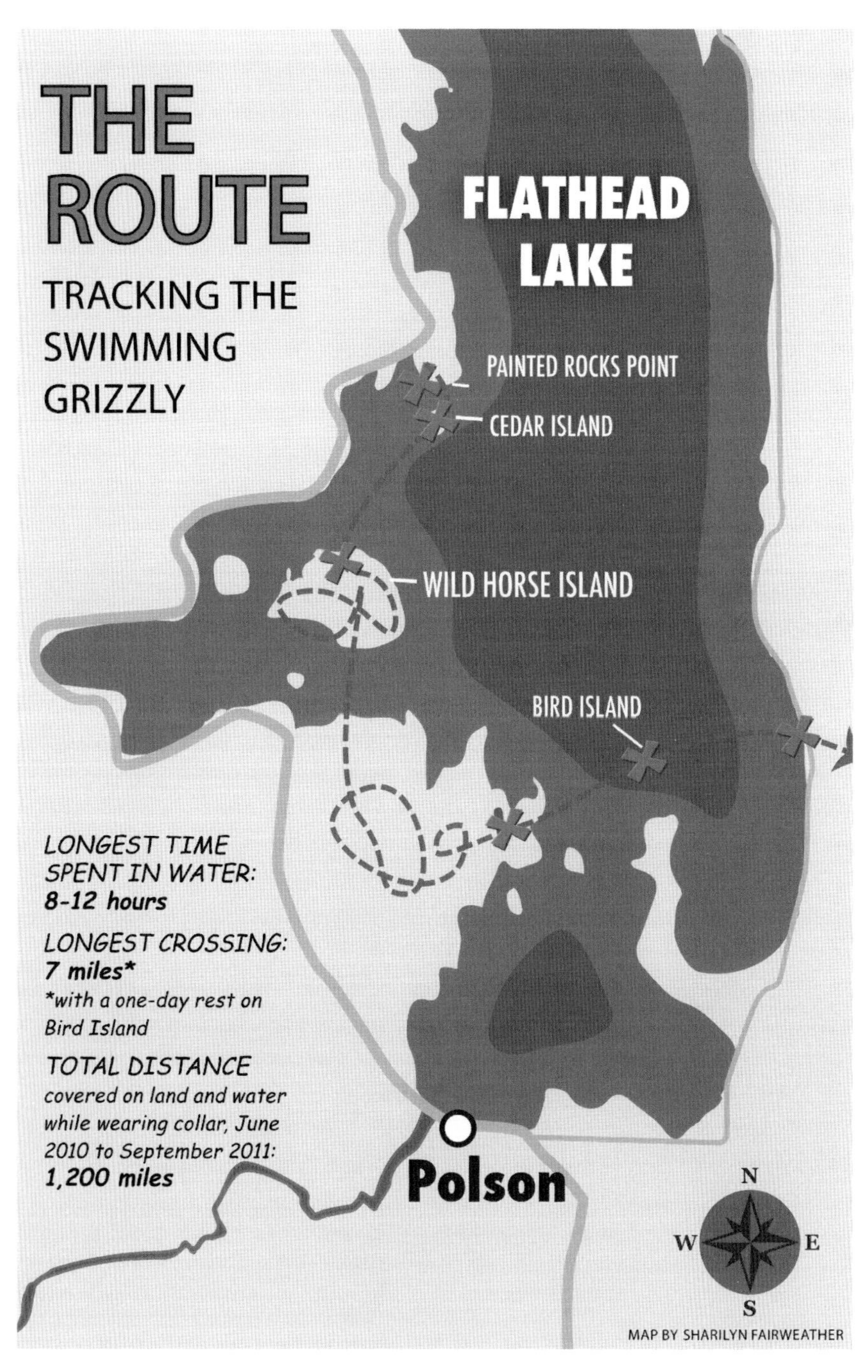

A map outlines the 2010 route of the grizzly bear who undertook a long swim in Flathead Lake. *Flathead Beacon map.*

"We thought because we don't have bears very often on the west side of the lake, we could put a collar on her," Rick Mace, a bear research biologist for Montana Fish, Wildlife & Parks, told a reporter from the *Daily Inter Lake* in 2011. The collar sent a signal of her location every four hours for more than a year. The bear dropped the collar near Swan Lake, where it was recovered by biologists. When the collar's electronic contents were downloaded, the bear's journey of more than 1,200 miles on land and in water was revealed.

Accustomed to widely wandering bears, Mace admitted the Lakeside Female surprised many folks. "Every bear is kind of cool when you look at the nooks and crannies where they've gone, but this is the first one that's done something like this."

Speaking to a reporter for the *Flathead Beacon* nearly four years after the grizzly's swim, bear researcher Lori Roberts offered a theory about the behavior of the Lakeside Female: "She was a young bear without many experiences in life yet. She was just out exploring." As for the swim, which apparently went undetected by humans, "it's pretty amazing. There had to be boats in the water, there are certainly a lot of houses. But nobody reported seeing her."

Researchers suspect that she spent up to twelve hours in water on the last leg of her swim to Bird Island and the east shore. "This the first time we've seen anything like that," Roberts admitted. "We have bears that swim Hungry Horse Reservoir but we've never had one swim Flathead Lake."

CHAPTER 2

A SPECIAL PLACE

How long is Flathead Lake? The answer depends on whom you ask. Some early historical accounts put the lake's length at 31 or 32 miles. Others say it's 28 miles, while the folks at the Flathead Lake Biological Station, experienced at measuring all sorts of things, say the lake is 27.3 miles long. But some of the folks with the most intense personal experiences with Flathead measure the lake not in miles but hours and minutes.

For Paul Stelter, who completed the first documented swim of the lake's length in 1986, it was twelve hours and forty-two minutes from Bird Point near Polson to Wayfarer's State Park near Bigfork, a distance of about twenty-five miles.

Almost two decades later, Ron Stevens of Whitefish covered the same route as Stelter in twelve hours, twenty-five minutes. It was his sixth attempt to swim the lake, and he celebrated his persistence with a post-swim trip to Burger Town, a landmark Bigfork eatery.

Emily von Jentzen, a Kalispell attorney, became the first woman to swim the lake in 2010, starting in Somers and finishing in Polson—a distance believed to be about twenty-eight miles—in eighteen and a half hours. By 2023, the lake's length had been conquered by at least seven people: Paul Stelter, 1986; Ron Stevens, 2005; Emily von Jentzen, 2010 and 2017; Sarah Thomas, 2015; Craig Lenning, 2015; John Cole, 2017; and Martyn Webster, 2023.

The big swim

Kalispell coach goes to Bigfork — the hard way

By JAMES E. LARCOMBE
Correspondent

BIGFORK — Winning what he described as a "battle," Kalispell swimmer Paul Stelter Tuesday overcame a stiff wind and exhaustion to become the first person to swim the length of Flathead Lake.

Stelter, 35, spent about 14 hours in bone-chilling water during his dawn-to-dusk, 25-mile swim. Sprinting for the lakeshore at Wayfarers, the trim Stelter was carried into Bigfork Bay by warm applause from more than 100 well-wishers.

Paul Stelter was the first to swim the length of Flathead Lake in 1986. *Missoulian.*

For some, swimming the largest freshwater lake west of the Mississippi River was akin to checking a box on an athletic to-do list amid an explosion in open-water swimming, ultramarathons and other endurance events around the globe. For others, the challenge was more personal.

In 1986, Stelter, a thirty-five-year-old Kalispell swim coach, was concerned about algae growth and related issues in the lake. A proposed local ban on phosphate laundry detergent, a factor in the algae growth, was making headlines. He hoped his swim would highlight the need to protect Flathead's water quality. "I love the lake," he told this author the day after completing his historic swim. "It was more of a tribute to the lake than anything." He added: "It's hellaciously hard on the mind and body. But I wouldn't want to do it anywhere else."

Stelter, an Illinois native, also confessed to having additional motivation. "It's been a personal goal ever since I moved here. How often in your lifetime do you have a chance to do something no one has ever done before?" Another admission while shivering on a dock after the swim: "I didn't feel like quitting, but it was hard. I never want to do it again."

As a toddler in 1986, von Jentzen was living in Washington State and had yet to complete a swim stroke. But in coming years, she took up swimming and eventually competed at the collegiate level. "I was a very clumsy child," she told this author in 2023. "I started swimming, and it turned out to be the one thing I could do better than my younger sister."

She started open-water swimming after moving to Missoula to attend law school and completed lengthy swims in a number of lakes in or near the Flathead Valley and in Glacier National Park. But when she dove into Flathead Lake in 2010, her longest open-water swim had been just four hours. After completing that swim in about eighteen and a half hours (she veered off course and swam closer to thirty-one miles than the expected twenty-eight), she was ready for more.

In 2011, von Jentzen tackled Lake Chelan, in north-central Washington, near where she grew up. She completed a fifty-five-mile swim, although it was a harrowing endeavor due to very rough water in the cold, narrow fjord-like lake. "It was like being in a washing machine," she recalled. "It's hard to negotiate hours and hours of waves like that." She spent two nights in the water, experienced hallucinations and weathered bouts of paranoia: "All the mental things; it was just insane." Completing the swim after thirty-six hours in the water, she suffered partial kidney failure, which she attributed to taking too much ibuprofen.

With her 2010 Flathead swim, von Jentzen raised nearly $10,000 for a young Missoula girl battling leukemia. When she swam Lake Chelan, donations went to a five-year-old with brain cancer. Come 2017, she had set her sights on Flathead again, the goal being an unprecedented down-and-back "double crossing" of the lake. Donations raised by the swim would help

Emily von Jentzen made Flathead Lake history in 2017 by swimming fifty-six miles from Somers to Polson and back. *Flathead Beacon photo.*

two more young children, one with cancer, the other with a heart disorder. All told, von Jentzen has raised more than $50,000 for young children and their families with her swims. "It's really cool to use something that I love to do something good," she told an interviewer.

In July 2017, von Jentzen waded into the lake at Somers at about six o'clock in the morning and headed for Polson, one stroke at a time. She was optimistic. She had trained for ten months, swimming as many as twenty hours per week. The hope was her early start would bring smooth conditions. But the lake didn't cooperate for long. By the time she reached the middle of the lake, "we had really big waves," she recalled. "It was like being on a swimming treadmill."

She was hours behind schedule and demoralized when she got to Polson at three o'clock in the morning. She admitted to having thoughts of abandoning the return leg. But John Cole, a Polson pediatrician who had trained with von Jentzen, was waiting on shore, where he had been for hours, along with a group of supporters. Cole was primed to make the return trip to Somers. Despite the long wait, "he was happy and engaging. He's like, 'Let's go,'" von Jentzen remembered.

Allowed no more than ten minutes on shore, von Jentzen huddled under a light towel, not wanting to get too warm for fear she wouldn't want to reenter the lake. "I ate two grilled cheese sandwiches and got back in." Pulling away from Polson in the dark, she recalled still having doubts, but Cole's presence propelled her. "I thought, 'He just got in; I can't get out.'"

Although he was fresh and a faster swimmer, Cole stayed close to von Jentzen, and the two plowed ahead. As the sky began to lighten, so did the mood. "The nights are the hardest," von Jentzen said. "The funnest part is when the sun starts to come up. Early in the morning is always when you have the flattest water."

While the pair encountered rough water on the return, it was more tolerable than on the first leg. Many followed the duo's northward journey electronically thanks to a GPS device tucked underneath von Jentzen's swim cap. When they emerged from the water at Somers, more than fifty people were on hand to greet them. Cole became the sixth person to swim the lake, while von Jentzen claimed the honor of being the only person to do a fifty-six-mile double crossing.

Speaking to this author in 2023, six years after her historic swim, von Jentzen, the mother of two young children and a busy partner in a Kalispell law firm, spoke wistfully of her long lake swims. "I do really think fondly on those experiences," she said. "It was a really fun time for me." But as a forty-year-old with plenty of other responsibilities, she had no plans for another long Flathead Lake adventure. Swimming, she said, "has been just a good, trusted friend. The parts that are going good, it's like euphoria. It's like a runner's high."

Despite the waves, the cold and the self-doubt served up by its waters, she holds Flathead Lake in warm regard. "The lake has always had a really special place in my heart," she said. "I don't feel like I could ever live anywhere else. I love it."

BIBLIOGRAPHY

Sources are listed in the order of their appearance in each chapter.

The Splendid Writings

Elrod, Morton J. "Pictured Rocks: Indian Writings on the Rock Cliffs of Flathead Lake, Montana, 1908." University of Montana Bulletin: Biological Series, 1901–10. https://scholarworks.umt.edu/umbiologicalseries/14.

White, Thain. *The Origin of Archaeological Specimens on the West Shore of Flathead Lake. A Composite, as Told to C.I. Malouf and T. White by Kutenai Informants, 1950*. Flathead Lake Lookout Museum, No. 2.

Keyser, James D. "Western Montana Rock Art: Images of Forgotten Dreams." *Montana: The Magazine of Western History* 42, no. 3 (Summer 1992): 58–69. https://www.jstor.org/stable/4519499.

———. Personal interview. August 2023.

The Lake's Big Boats

Braunberger, A.B. "Flathead Lake Steamboats." Self-published research paper, 1965.

Whitefish (MT) Pilot. "Freight Steamer Big Fork Sinks." January 6, 1910.

Franz, Justin. "The Search for the Kee-O-Mee." *Flathead Beacon* (Kalispell, MT), November 19, 2016.

Fugleberg, Paul. "Steamboat Skippers Battled a Tempestuous Flathead Lake." *Missoulian* (Missoula, MT), December 23, 2013.

White, Thain. *Vessels & History of Flathead Lake*. Publication No. 15, Flathead Lake Lookout Museum. Date unknown.

Jones, Dennis O. Personal interview. August 2023.

Lake Shore Sentinel (Polson, MT). December 10, 1909.

Fugleberg, Paul. "MS Flathead, Pride of Lake's Boating Fleet, Is Destroyed by Flames in Lake Mishap." *Flathead Courier* (Polson, MT), June 22, 1961.

The Durable Helena

Braunberger, A.B. "Flathead Lake Steamboats." Self-published research paper, 1965.
Kehoe, Jack. "Steamboating on Flathead Lake." In *Historical Information Concerning the Upper Flathead Country.* Kalispell, MT: Trippets Printing, 1956.
Kehoe, Leslie. Personal interview. June 2023.
Kehoe, James J. Letter to editor. *Daily Inter Lake* (Kalispell, MT), date unknown.
Chaix, Jaix. "Kehoe's Agate Shop." *Flathead Beacon* (Kalispell, MT), October 14, 2014.
Reiss, Mackenzie. "One Family Is Heart of Kehoe's Agate Shop for Nearly 90 Years." *Daily Inter Lake* (Kalispell, MT), March 1, 2020.

From Paths to Pavement

Elwood, Henry. *Kalispell, Montana and the Upper Flathead Valley.* Kalispell, MT: Thomas Printing, 1989.
White, Thain. *Medicine Rock.* Lakeside, MT: Flathead Lake Lookout Museum, September 1960.
Hanson, Eric. Personal interview. August 2023.
———. Communication with Doug Johns. 2011.
Axline, Jon. "Building Permanent and Substantial Roads: Prison Labor on Montana Highways, 1910–1925." *Montana: The Magazine of Western History* 62, no. 2 (Summer 2012): 59–66, 95–96. https://www.jstor.org/stable/24414628.
———. Personal interview. July 2022.
Jones, Dennis O. Personal interview. July–August 2023.

A Company Town

Elwood, Henry. *Somers, Montana: A Company Town.* Somers, MT: Somers Company Town Project, 2015.
Heslewood, Fred W. "The Somers Strike: Stay Away from Somers, Mt." *Industrial Worker* (Spokane, WA), July 29, 1909.
Hellen, Ruth. Personal interview. Summer 2018.
Ruby, Dave. Personal interview. Summer 2018.
Wondrow, Carolyn. Personal interview. Summer 2018.

The Mansion on the Hill

Elwood, Henry. *Somers, Montana: A Company Town.* Somers, MT: Somers Company Town Project, 2015.

Chaix, Jaix. "The O'Brien Mansion." *Flathead Beacon* (Kalispell, MT), November 24, 2014.
Daily Inter Lake (Kalispell, MT). "Woman Sentenced in Somers Mansion Fraud Case." March 12, 2015.
Serbin, Bret Ann. "A New Chapter Begins for Historic Somers Mansion." *Daily Inter Lake* (Kalispell, MT), May 3, 2020.
Franz, Justin. "Saving the Mansion on the Hill." *Flathead Beacon* (Kalispell, MT), September 9, 2021.
Morton, Jasmine. Personal interview. Spring 2023.
Jones, Dennis O. Personal interview. August 2023.
Somers Mansion. "History of the Somers Mansion." www.somersmansion.com/history.

The Journey of the Paul Bunyan

Braunberger, A.B. "Flathead Lake Steamboats." Self-published research paper, 1965.
Elwood, Henry. *Somers, Montana: A Company Town.* Somers, MT: Somers Company Town Project, 2015.
Devlin, Vince. "Towboat in Trouble: Beloved Paul Bunyan Logging Ship Needs a Roof Over Its Head." *Missoulian* (Missoula, MT), May 3, 2010.
Larcombe, James E. "Lake Landmark Gets New Berth." *Missoulian* (Missoula, MT), May 21, 1987.
Daily Inter Lake (Kalispell, MT). "Californian Buys Flathead Lake Lookout." June 20, 1971.
Mangels, Gil. Personal interview. Fall 2021.

Building the Dam

Missoulian (Missoula, MT). "Montana Power Gets Lease." May 20, 1930.
Kirk, Cecil H. *A History of the Montana Power Company*. Donn Kirk, 2008.
Missoulian (Missoula, MT). "Seven Bodies under Big Earth Slide in Canyon." March 3, 1937.
Great Falls (MT) Tribune. "Seven Killed, 3 Hurt by Rock Avalanche at Polson Dam Project." March 3, 1937.
Missoulian (Missoula, MT). "All Flathead Joins in Dedication," August 7, 1938.
Montana Standard (Butte, MT). "Red Man and White Man Join in Dam Dedication." August 7, 1938.
Lipscomb, Brian. Personal interview, 2021.
Flathead Courier (Polson, MT). Various issues, 1930–38.
Smith, Thompson. "A Brief History of Kerr Dam and the Reservation." Essay based on the script for *The Place of Falling Waters*, a documentary film by Roy BigCrane and Thompson Smith.

Thain White and the Informants

White, Thain. *Vessels & History of Flathead Lake.* Flathead Lake Lookout Museum Educational Publication No. 15. Date unknown.
Daily Inter Lake (Kalispell, MT). Thain White obituary. April 30, 1999.
Thain White Papers. Archives and Special Collections, Mansfield Library, University of Montana, Missoula, Montana.
Jones, Dennis O. Personal interview. July–August 2023.
Carling Malouf Oral History Project. 2004. Archives and Special Collections, Mansfield Library, University of Montana, Missoula, Montana.
White, Ruth. Personal interview. September 2023.

The Flathead Lake Fight of 1943

Daily Inter Lake (Kalispell, MT). June 2–4, 1943.
Rockwood, Forrest, and T.B. Moore. Letter to editor. *Daily Interlake* (Kalispell, MT), May 22, 1943.
Associated Press. "Flathead Plan Is Called Least Evil." June 2, 1943.
Tabish, Dillon. "The Flathead Lake Fight of 1943." *Flathead Beacon* (Kalispell, MT), April 19, 2016
Missoulian (Missoula, MT). "Flathead in State's Rights Stand." June 4, 1943.
Daily Inter Lake (Kalispell, MT). "Raver Says Lake Plan Given Up." July 23, 1943.
Montana Standard (Butte, MT). "FDR Interested in Flathead Case." August 26, 1943.
Monoson, Ted. "Mansfield's Story: New Biography Remembers Montana Senator Who Asked to Be Forgotten." *Missoulian* (Missoula, MT), October 7, 2003.
Oberdorfer, Don. *Senator Mansfield: The Extraordinary Life of a Great American Statesman.* Washington, D.C.: Smithsonian Books, 2003.

War Comes to the Lake

Mikesh, Robert C. *Japan's World War II Balloon Bomb Attacks on North America.* Washington, D.C.: Smithsonian Institution Press, 1973.
Tanglen, Larry. "Terror Floated Over Montana: Japanese Balloon Bombs 1944–1945." *Montana: The Magazine of Western History* 52, no. 4 (Winter 2002): 76–79. https://www.jstor.org/stable/4520467.
Western News (Libby, MT). "Jap Balloon Found in Timber." December 21, 1944.
Salt Lake Telegram (Salt Lake City, UT). "Jap Balloon Bomb Found in Montana Forest." December 19, 1944. Associated Press article.
Axline, Jon. "Balloon Bombs: Several Windship Weapons Launched by Japan in WWII Landed in Montana." *Independent Record* (Helena, MT), February 26, 2012.

The Lake Watchdog

Lamb, Janelle Leader. "A Century of Service: The Flathead Lake Biological Station Celebrates One Hundred Years of Education, Research and Outreach." 1999.

Bansak, Tom. Personal interviews. June 2016 and August 2023.

Elser, Jim. Personal interview. June 2016.

Stanford, Jack. Personal interview. July 2016

Malison, Rachel. Personal interview. September 2019.

Matson, Phil. Personal interview. September 2019.

The Relentless Morton Elrod

Dennison, George M. *Montana's Pioneer Naturalist: Morton J. Elrod.* Norman: University of Oklahoma Press, 2016.

Lamb, Janelle Leader. "A Century of Service: The Flathead Lake Biological Station Celebrates One Hundred Years of Education, Research and Outreach." 1999.

Elrod, Morton J. "Pictured Rocks: Indian Writings on the Rock Cliffs of Flathead Lake, Montana, 1908." University of Montana Bulletin: Biological Series, 1901–10. https://scholarworks.umt.edu/umbiologicalseries/14.

Missoulian (Missoula, MT). "Dr. Elrod Dies in Sleep at Age 89." January 19, 1953.

Tom Bansak. Personal interview. August 2023

Larcombe, James E. "Review: Montana's Pioneer Naturalist, Morton J. Elrod." *Montana Magazine*, 2017.

Elrod, Morton J. "Some of the Last Free Government Homestead Land: The Flathead Reservation." Northern Pacific Railway, 1909.

A Love of the Lake

Reksten, Patty. "Personality Profile: Dr. Jessie M. Bierman." *Daily Inter Lake* (Kalispell, MT), November 28, 1971.

Schwennesen, Don. "Jessie Bierman: A Health-Care Pioneer." *Missoulian* (Missoula, MT), July 26, 1982.

SF Gate (San Francisco, CA). Jessie Bierman obituary. September 7, 1996.

Lamb, Janelle Leader. "A Pioneering Spirit—Jessie Bierman, Patroness of the Flathead Lake Biological Station." *Montanan* (University of Montana) 16, no. 3 (Summer 1999): 12–13.

Bansak, Tom. Personal interview. August 2023.

On Shaky Ground

Daily Inter Lake (Kalispell, MT). "Moderate Quake Shakes Valley." April 1, 1952.

Stickney, Mike. Personal interviews. Summer 2020.

Macknicki, Jim. "Quake Rocks Kalispell; No Damage or Injuries." *Daily Inter Lake* (Kalispell, MT), February 4, 1975.

Franz, Justin. "Geologist: Recent Shakes Reminder That Montana Is Earthquake Country." *Flathead Beacon* (Kalispell, MT), November 17, 2014.

The Wild Story of Wild Horse Island

McCurdy, Edward. "Wild Horse Island: Yesterday, Today and Tomorrow." Self-published research paper, 1975.

Di Salvatore, Bryan. *Flathead Lake On My Mind.* Pediment Publishing, 2015.

The Story of Wild Horse Island. Kootenai Committee of the Confederated Salish and Kootenai Tribes, 1982.

Mercer, John. Personal interview. Spring 2021.

Grout, Amy. Personal interview. Spring 2021.

MacDonald, Paddy. Personal interview. Winter 2021.

Bansak, Tom. Personal interview. Spring 2021

Anderson, Marge. "Wild Horse Island—Enchanting." *Missoulian* (Missoula, MT), August 13, 1968.

Scott, Tristan. "Wild Horse Island Yields New World Record Bighorn Ram." *Flathead Beacon* (Kalispell, MT), February 9, 2018.

———. "3 Mountain Lions Killed on Wild Horse Island." *Flathead Beacon* (Kalispell, MT), March 5, 2022.

The Almost-Forgotten Life of Herman Schnitzmeyer

Fugleberg, Paul. "Talented, Eccentric Schnitzmeyer Was a Very Special Person." *Ravalli Republic* (Hamilton, MT), March 21, 2016.

———. *Schnitzmeyer: Homestead Era Photographic Artist.* Self-published, 1992.

Kellogg, Dennis. "Stories and Stones: Herman Schnitzmeyer." Video. MCAT Community Media, 2006. https://www.youtube.com/watch?v=ykQ1M2PzzWY.

Kellogg Dennis. Personal interview. August 2023.

Stromnes, John. "Almost Famous Homestead-Era Photographer Never Got His Due but Now His Artistic Sense Is Being Recognized." *Missoulian* (Missoula, MT), July 30, 2006.

Daily Inter Lake (Kalispell, MT). "Maroney Made Spectacular Flight Sunday." July 28, 1913.

Masons and Merit Badges

New Age Magazine. "Masonry at Work, Missoula, MT." January 1927.

Apple, Bernice. "Polson Museum Notes." *Flathead Courier* (Polson, MT), March 13, 1980.

Stromnes, John. "Scout Island." *Missoulian* (Missoula, MT), July 27, 2003.

Orange County (CA) Register. Fred Cox obituary. August 31, 2014.

Gadbow, Daryl. "Melita Island Deal Done." *Missoulian* (Missoula, MT), February 9, 2005.

Mann, Jim. "Scouts to Buy Melita Island." *Daily Inter Lake* (Kalispell, MT), December 31, 2004.

Fish, Andrew. "Local Island Becoming National Attraction." *Lake County Leader* (Polson, MT), July 31, 2008.

The Mysterious Creature

Missoulian (Missoula, MT). "Flathead Lake Whopper Hits 7½ Feet." May 29, 1955.

Daily Inter Lake (Kalispell, MT). "Area Man Lands 7½-foot Flathead Lake Sturgeon." May 29, 1955.

Flathead Courier (Polson, MT). June 2, 1955.

Allen, A.E. "Fisherman Hooks Flathead Lake Monster." *Great Falls (MT) Tribune*. June 12, 1955.

Daily Inter Lake (Kalispell, MT). "Griffith Says Reports of Big Catch Overdone." June 2, 1955.

———. "MSU Scientist Continues Study of Lake "Baby" Monster." June 2, 1955

———. "Judge Upholds Jury Verdict on Ownership of Big Fish." June 23, 1957.

Brunson, Royal B., and Daniel G. Block. *The First Report of the White Sturgeon from Flathead Lake Montana.* Proceedings of the Montana Academy of Sciences, 1957.

Montana Supreme Court. *Griffith v. McAlear, No. 9844.* December 4, 1959.

New York Times. "Montana Has a Loch Ness Monster." June 13, 1965.

Stromnes, John. "Sounds Fishy…or Does It?" *Missoulian* (Missoula, MT), March 27, 2003.

Fugleberg, Paul. "Flathead Lake Sturgeon Catch Still Controversial." *Missoulian* (Missoula, MT), April 5, 2015.

Devlin, Vince. "Lake Creatures Saves Tot's Life: Flathead Monster Stories Go Back More Than a Century." *Missoulian* (Missoula, MT), January 1, 2017.

Daily Inter Lake (Kalispell, MT). Delano "Laney" Anthony Hanzel obituary. October 1, 2022.

Devlin, Hilary. Personal interview. October 2022.

Dunwell, Karen. Personal interview. December 2022.

Bansak, Tom. Email interview. January 2023.

Cold, Dark and Lonesome

Missoulian (Missoula, MT). "Jet Believed Down in Flathead Lake." March 22, 1960.

Fugleberg, Paul. "Crew Locates Debris from Jet." *Missoulian* (Missoula, MT), March 23, 1960.

Missoulian (Missoula, MT). "Divers to Probe Navy Jet Crash." March 24, 1960.

———. "Not Feasible to Raise Navy Jet." March 27, 1960.
———. "Eaheart Search in Flathead Lake Is Ended." March 30, 1960.
Larrson, Robert C. "Flathead Dock Builder Is Ready to Sail." *Missoulian* (Missoula, MT), January 24, 1965.
United States Navy. *Aircraft Accident Report*. OPNAV 3750-1. April 6, 1950.
Stang, John. "Part of Missing Navy Jet Located." *Daily Inter Lake* (Kalispell, MT), May 3, 2006.
Stang, John. "Pilot's Remains Found." *Daily Inter Lake* (Kalispell, MT), May 9, 2006.
Stromnes, John. "Side-Scan Sonar Locates Jet in Flathead Lake." *Missoulian* (Missoula, MT), May 3, 2006.
Devlin, Vince. "Searchers Locate Pilot's Body in Flathead Lake." *Missoulian* (Missoula, MT), May 9, 2006.
Gruley, Bryan. "A Pilot Reaches Deep to Plumb the Mystery of Another's Crash." *Wall Street Journal*. May 23, 2006.
Daily Inter Lake (Kalispell, MT). Viola Lewis obituary. January 23, 2023.
Horner, Doug. "Bringing Up the Bodies: The Retired Couple Who Find Drowning Victims." *Guardian*, January 16, 2020.
Sentinel (yearbook). Missoula, MT: University of Montana, 1949.
Bigfork History Project. Scout Jessop interview with Vi Lewis. 2019.
Behar, Michael. "Cold Case: A New Team Sets Out to Solve Old Disappearances." *Smithsonian Air and Space*, September 2010.

The Explosive Mystery of Medicine Rock

White, Thain. *Medicine Rock*. Lakeside, MT: Flathead Lake Lookout Museum, September 1960.
Hanson, Eric. Personal interview. August 2023.
Weaver, Rick. Personal interview. August 2023.
Jones, Dennis O. Personal interview. July 2023.

The Night the Music Died

Jones, Dennis. "Flathead Crash Kills All Aboard." *Missoulian* (Missoula, MT), July 5, 1987.
———. "Crash Kills Montana Band Members." *Missoulian* (Missoula, MT), July 6, 1987.
———. "State's Worst Crash Probed." *Missoulian* (Missoula, MT), July 7, 1987.
Downey, Janice. "A Band That Captured the Spirit of Montana." *Missoulian* (Missoula, MT), July 6, 1987.
Daily Inter Lake (Kalispell, MT). July 5–8, 1987.
Melton, Wayne. "News of Crash Stuns Colleagues." *Reno* (NV) *Gazette-Journal*, July 6, 1987.
National Transportation Safety Board. *Aviation Investigation Final Report, Accident—DEN87MA168, July 4, 1987*. Report completed February 24, 1989.

The Adventures of the Lakeside Female

Landers, Rich. "Grizzly Bear Amazes Researchers with Long Swims in Flathead Lake." *Spokesman-Review* (Spokane, WA), September 19, 2011.
Missoulian (Missoula, MT). "Young Grizzly Bear a Champion Swimmer." September 19, 2011.
Scott, Tristan. "The Bobbing Bruin." *Flathead Beacon* (Kalispell, MT), April 7, 2015.

A Special Place

Larcombe, James E. "The Big Swim: Kalispell Coach Goes to Bigfork the Hard Way." *Missoulian* (Missoula, MT), August 20, 1986.
———. "Super Swim." *Missoulian* (Missoula, MT), August 21, 1986.
Stromnes, John. "Kalispell Man Second Ever to Swim the Length of Flathead Lake." *Missoulian* (Missoula, MT), August 21, 2005.
Franz, Justin. "A History-Making Swim." *Flathead Beacon* (Kalispell, MT), July 31, 2017.
FLOW (Flathead Lake Open Water) Swimmers. https://flowswimmers.com.
Von Jentzen, Emily. Personal interview. September 2023.

INDEX

D

E

F

G

H

I

J

K

L

M

N

O

P

Q

R

S

T

U

V

W

Y

Z

ABOUT THE AUTHOR

James E. "Butch" Larcombe is a fourth-generation Montanan who grew up in Malta, a small town on the state's Hi-Line. From his earliest years, his family spent time at Flathead Lake. A graduate of the University of Montana, Larcombe has worked as a social studies teacher in the Bitterroot Valley and later as a newspaper reporter and editor in Missoula, the Flathead Valley, Great Falls and Helena. He also was the editor and general manager of *Montana Magazine* for six years before joining the communications department at NorthWestern Energy.

Along with thousands of newspaper articles, Larcombe's work has appeared in *Montana Quarterly* magazine, *Montana Outdoors* and in the *Flathead Beacon* and its quarterly magazine, *Flathead Living*. He is the author of two nonfiction books: *Golden Kilowatts: Water Power and the Early Growth of Montana* and *Montana Disasters: True Stories of Treasure State Tragedies and Triumphs*. He lives near Woods Bay, Montana.